F-86 SABRE in action

by Larry Davis

Color by Don Greer
Illustrated by Tom Tullis

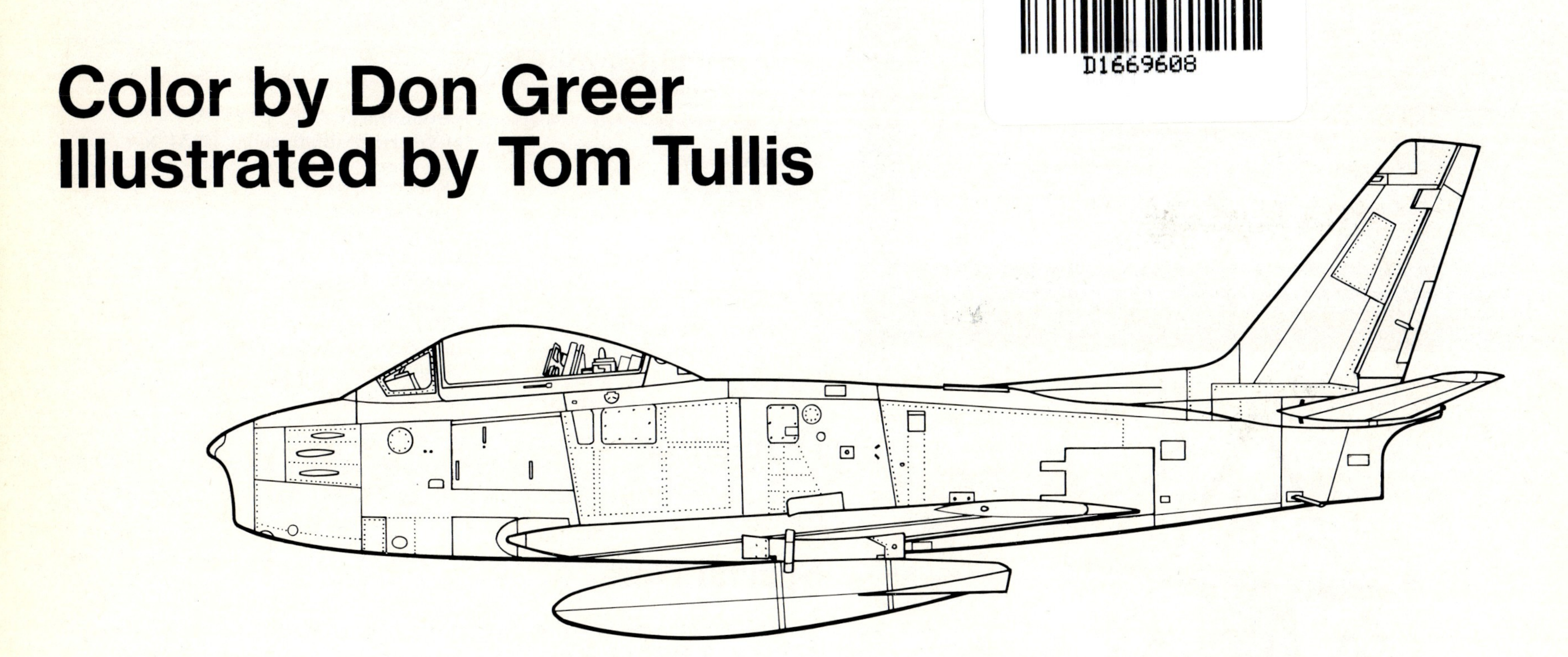

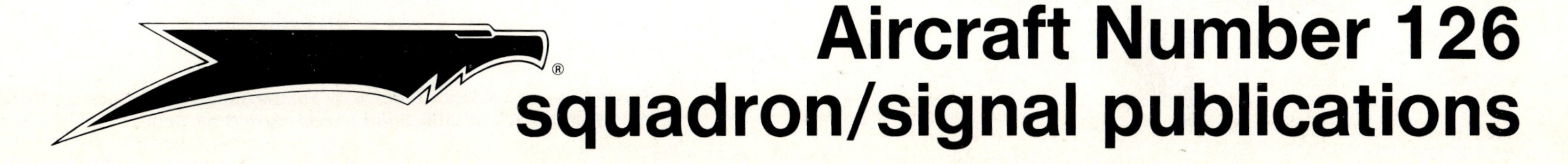

Aircraft Number 126
squadron/signal publications

CAPT James Jabara scored his fifth and sixth MiG kills on 20 May 1951 to become the world's first Jet Ace. He was able to score both kills even though he was not able to drop his wing tanks.

ISBN 0-89747-282-9

If you have any photographs of the aircraft, armor, soldiers or ships of any nation, particularly wartime snapshots, why not share them with us and help make Squadron/Signal's books all the more interesting and complete in the future. Any photograph sent to us will be copied and the original returned. The donor will be fully credited for any photos used. Please send them to:

Squadron/Signal Publications, Inc.
1115 Crowley Drive.
Carrollton, TX 75011-5010.

Acknowledgements

Air Force Museum
Peter Bowers
Drury Callahan
Robert F. Dorr
Jeffery Ethell
Marty Isham
JEM Aviation Slides
Robert L McKinney
Dave McLaren
North American Aviation
Royal Netherlands Air Force
Jim Sullivan
Nick Waters
Brian Baker
COL Henry Buttlemann, USAF Ret.
Irv Clark
Robert Esposito
Leo Fournier
Italian Air Force
Oscar Lind
George McKay
David Menard
Merle Olmsted
Sabre Pilots Association
United States Air Force

Dedication

For my good friend David Menard - who always TRIES to keep me straight. SABRES FOREVER!

Special Note

Former Sabre pilots interested in an organization dedicated to preserving the history of the F-86 should contact:

The F-86 Sabre Pilots Association
P.O. Box 97951
Las Vegas, NV 89193

This North American Aviation F-86F Sabre of the 336th Fighter Interceptor Squadron carries the Yellow and Black identification bands carried by Sabres during the Korean War. (E. Sommerich)

USAF
U.S. AIR FORCE
12894
FU-894

Introduction

SABREJET! — That word instantly brings to mind thoughts of silver swept wing jets locked in mortal combat in the skies over Korea. But the North American Aviation (NAA) F-86 Sabre was much more than that. Not only was it the best clear-air fighter in the world at that time, the basic design would evolve into the only production single seat all-weather jet interceptor, an atomic capable fighter-bomber and a carrier-based fleet interceptor and fighter-bomber for the Navy. The F-86 series was built by many different manufacturers throughout the Free World and served to counter the Soviet threat during the 1950s.

The F-86 Sabre was just one of the superb designs to come from North American Aviation. During the 1940s and 1950s North American literally "owned" the military aircraft sales market, not just in America, but throughout the Free World. This aviation giant began life as a holding company in the early 1930s with no designs of its own. Originally founded in 1928, the company was reorganized in 1933 with James H. "Dutch" Kindelberger as President. Under Kindelberger, NAA entered the military aviation business.

The first NAA design was the O-47 observation aircraft. This was followed by a single engine, monoplane basic trainer, the BT-9 which evolved into the greatest trainer aircraft of the Second World War — the T-6 Texan. The BT-9/T-6 contract made it possible for NAA to move to a new North American Aviation plant in Inglewood, California in January of 1936. The 159,000 square foot factory was located at the south edge of Mines Field (Los Angeles International Airport).

During the Second World War, NAA produced two of the war's finest combat aircraft. The first was the famed B-25 Mitchell and the second aircraft was the "most perfect pursuit plane in existence" — the P-51 Mustang. At the peak of wartime operations, NAA employed over 91,000 people. The factory had grown into two huge complexes with over 8,573,835 square feet of floor space. Between 1 January 1939 and 30 September 1945, NAA built 42,683 military aircraft including 15,603 P-51s, over 15,400 T-6s and 9,817 B-25s (some 14 percent of all U.S. aircraft production).

An incident midway through the war made the Mustang obsolete, along with every other combat aircraft in existence. That incident was the operational debut of the German jet-powered Messerschmitt Me-262. With two Jumo jet engines, the Me-262 had a speed of over 550 mph, some 125 mph faster than the best Allied fighters. By early Fall 1944, Me-262s were defending the Reich, taking the Allies by surprise.

The rush to catch up saw jet aircraft design proposals come in from every U.S. company. Bell Aircraft was the first to have a jet fighter aircraft flying. The P-59 Airacomet, however, was not a successful combat aircraft: it was not as fast as the Mustang or Thunderbolt and could not hope to match the Me-262. Lockheed and Gloster fielded jet fighters and both the Lockheed P-80 and the Gloster Meteor did see some combat in the Second World War, but neither was a counter to the Me-262. Three other companies submitted design proposals to the War Department: Republic (XP-84), Grumman (XF9F) and North American (XP-86).

NAA had been interested in jet powered aircraft for some time, with the Confidential Design Group making an initial proposal for a jet fighter in early 1943. Ed Schmued had one design based on a piston/turbojet powered P-51 with a radical forward swept wing. One of the design studies from North American was selected by the Navy as its first jet fighter aircraft, the XFJ-1 Fury. The XFJ-1 used a circular fuselage that enclosed a GE J35 jet engine. The intake was in the nose with a straight through duct to the engine, rather than using side inlets such as those found on the P-80 and XF9F. The wing and tail were similar to the P-51 Mustang and armament was six .50 caliber machine guns. The contract for the first XFJ-1 Fury was let on 1 January 1945.

The original XP-86 design was quite similar to the XFJ-1 Fury. The XP-86 fuselage was more slender and longer than the Fury, but it retained the wing and tail surfaces of the XFJ-1. Both the XP-86 and the XFJ-1 were powered by a 4,000 lbst General Electric J35 axial flow turbojet. The straight wing and the low thrust of the J35 restricted the XP-86 to a top speed of some 575 mph — far below what the USAAF requirement called for.

Project drawings were for a proposed compound piston/jet variant of the P-51 Mustang with a forward swept wing. (NAA)

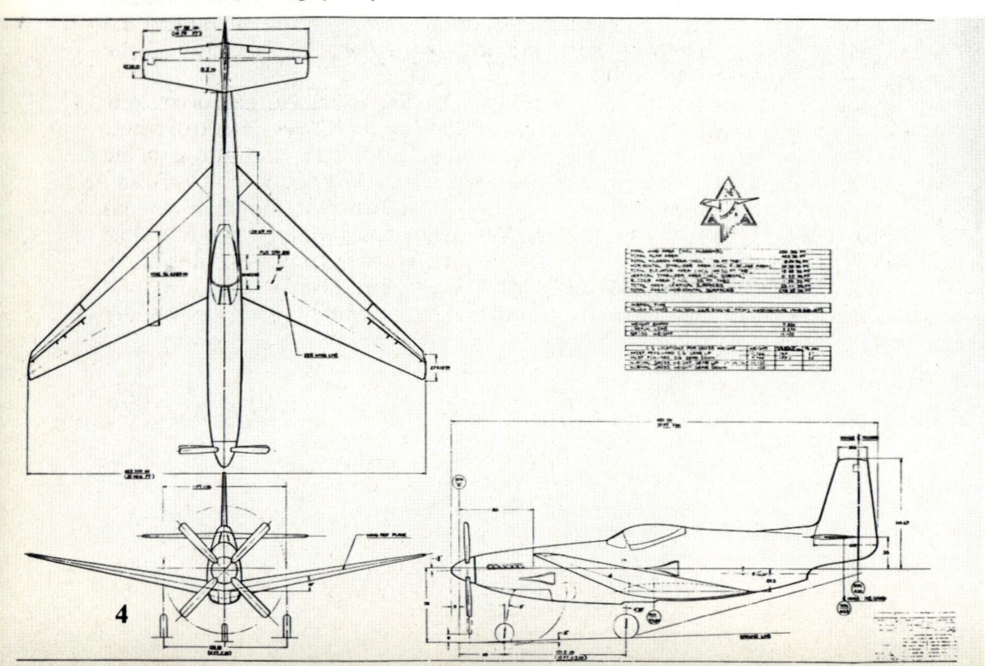

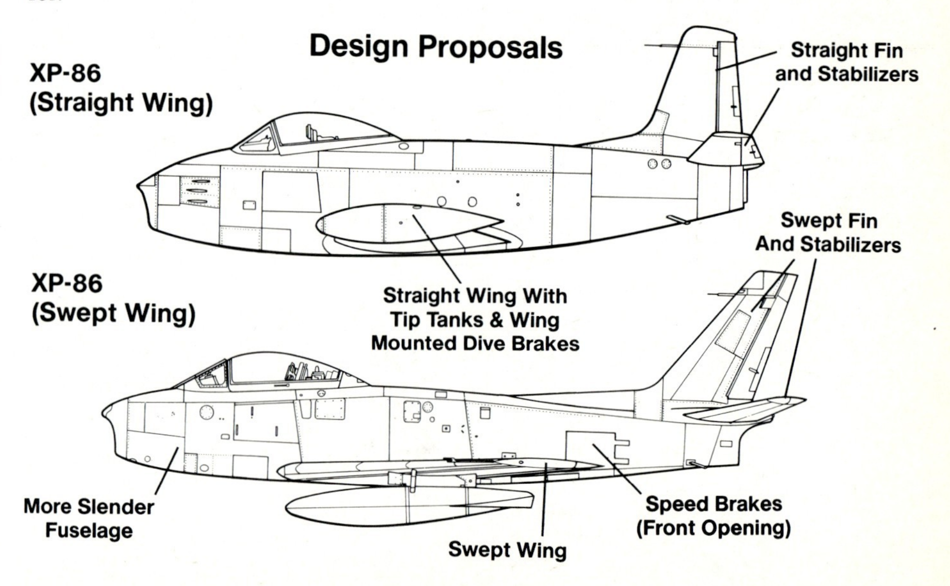

The end of the Second World War brought an immediate (and controversial) influx of jet technology from Germany. As rapidly as advances were made in engine technology, however, the speed potential seemed to stabilize at about 600 mph. Something was needed to bring airframe technology in line with the power plants. Captured German documents, translated by Larry Green, NAA's Head of Design Aerodynamics, including transonic wind tunnel data, clearly showed the speed benefit of swept wing surfaces.

In September of 1945, NAA engineer Harrison Storms studied the results of Green's research and thought it might be incorporated into the XP-86 proposal. The German wind tunnel tests revealed compressibility drag was delayed through the use of swept wings. Storms' swept wing XP-86 was estimated to be about 75 mph faster than the straight wing XP-86. There were problems with the use of a swept wing including a loss in low speed handling characteristics, but a further study of German research revealed that movable leading edge devices, called slats, that extended into the airstream improved low speed handling.

Storms' new wing with its 35 degree sweep and full span leading edge slats was fitted to the redesigned XP-86 model and tested in NAAs low speed wind tunnel. The swept wing exhibited satisfactory stall characteristics when fitted with the leading edge slats and gave the XP-86 an estimated top speed of over 650 mph. As a result, the USAAF authorized three prototypes in November of 1945.

The first XP-86 prototype (45-59597) took slightly over eighteen months to complete and rolled out at NAA's Inglewood plant on 8 August 1947. The prototype emerged as a sleek low wing aircraft with swept back wings and tail surfaces, with little resemblance to its contemporaries, such as the P-80, P-84 and F9F. Not only was the airframe a radical departure from any other aircraft then in production, the XP-86 also incorporated two major structural innovations. First was the use of tapered skinning and the second was the use of a double skin structure rather than a conventional rib and stringer construction. This provided added strength and allowed enough open area in the wing for fuel tanks.

The XP-86 looked fast — and it lived up to its looks. After extensive ground testing, NAA test pilot George "Wheaties" Welch lifted the XP-86 off the dry lake bed at Muroc on 1 October 1947. Flight tests revealed that the XP-86 was all that the Air Force had hoped for. The greatest advantage of the swept wing was its low drag at high Mach numbers.

The swept wing XP-86 rolled out on 8 August 1947 and NAA test pilot, George "Wheaties" Welch made the first flight with the XP-86 from Muroc Dry Lake on 1 October. The prototype carried no armament. (NAA)

The first XP-86 prototype during a test flight over California with NAA test pilot "Wheaties" Welch at the controls. Welch put the XP-86 through the sound barrier on 25 April 1948 - the first production airplane in history to go supersonic. (NAA)

This meant that thrust that was needed to reach these speeds with a straight wing aircraft could now be used for combat maneuvering. It was found that the swept wing actually had a higher lift factor than a larger straight wing design at high Mach numbers. This meant that a Sabre pilot could keep the Mach number high in a diving turn where lift is usually a limiting factor with a straight wing. The XP-86 was designed with a set of speed brakes in the rear fuselage, two on the fuselage sides and one, large brake under the fuselage. These brakes could be opened in any attitude and speed, including at or above Mach One. No other fighter had such a braking system.

Tests showed that the XP-86 was 75 mph faster than anything else in service. The XP-86 wasn't just a fast aircraft. The leading edge slats could be used to lessen turn rate and gain an advantage over an enemy. It was a "pilots airplane." If the XP-86 went out of control, the pilot need only to release the stick and the XP-86 would stabilize, if you had the altitude. One of the things pilots liked about the airplane was the great vision through the 360 degree bubble canopy.

Low speed handling, especially during takeoff and landing was still an area of concern, even with full extension of the leading edge slats. If the wing lost lift, no amount of power could keep the XP-86 airborne. The aircraft would perform what was known as the "Sabre dance," during which the aircraft would flounder in the air, waving back and forth on its exhaust plume, usually with fatal results. The answer was more power and North American re-engined one of the three XP-86 prototypes with the new 5,200 lbst J47-GE-3 power plant (which was intended for production aircraft) and it was this XP-86 that "Wheaties" Welch put through the sound barrier on 25 April 1948. With the successful completion of the test program, the Air Force ordered the aircraft into production under the designation F-86A Sabre.

Development

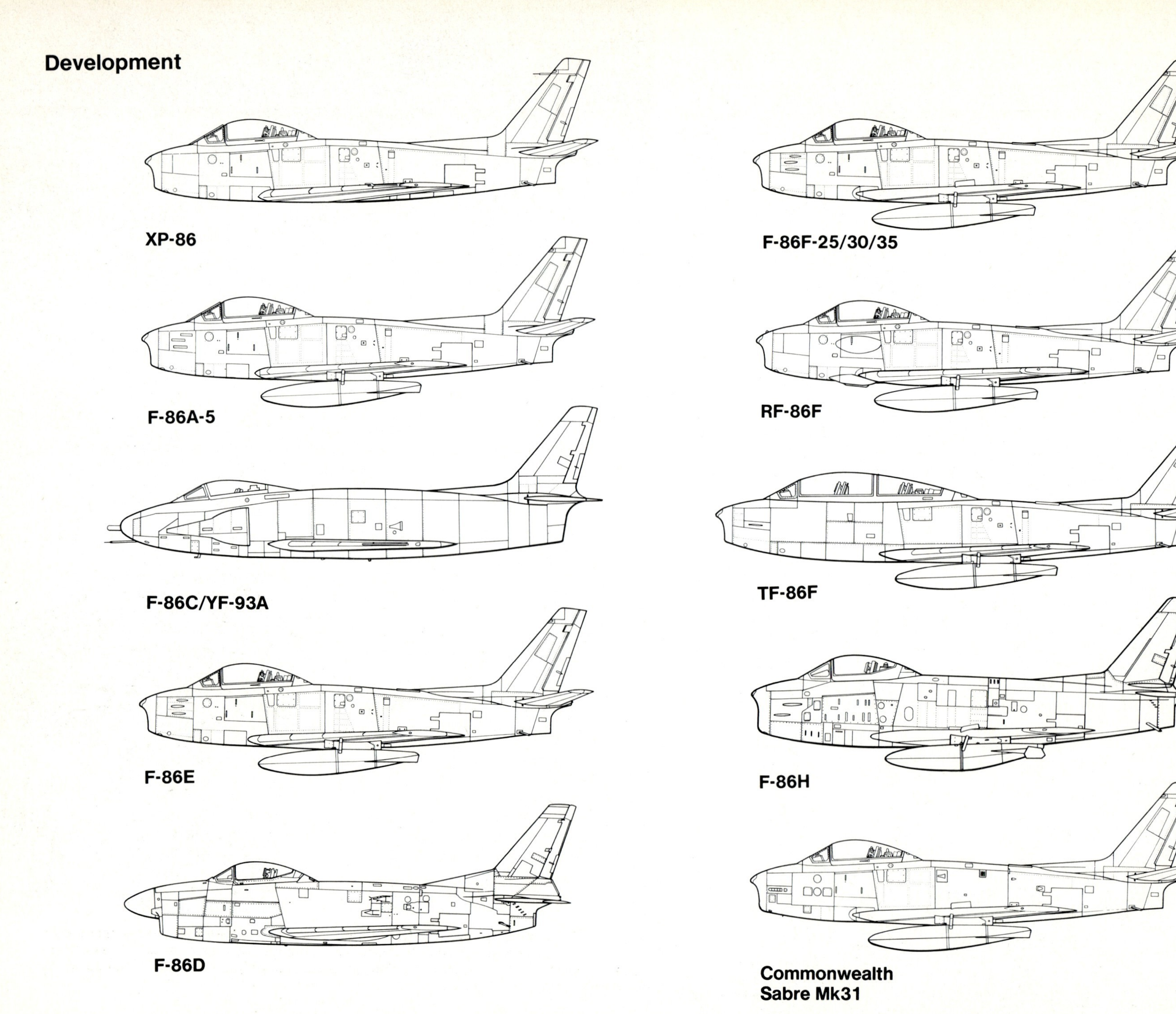

F-86A

The installation of the J47-GE-1 engine was the main difference between the XP-86 prototype and the F-86A production aircraft. The J47-GE-1 engine was rated at 5,200 lbst, compared with the 4,000 lbst available with the J35 engines used in the XP-86s. Installation of the J47 brought the performance of the F-86A-1 into the transonic region, with a top speed of over 670 mph. MAJ Robert L. Johnson exhibited this performance when he set a new World Speed Record of 670.98 mph in a production F-86A-1 on 15 September 1948.

Externally the production F-86A differed only slightly from the XP-86 prototypes. The F-86A-1 was armed with six .50 caliber machine guns which had not been installed on the XP-86. The guns were in three-gun packs installed on each side of the fuselage nose. The gun ports were covered with electrically operated doors for aerodynamic streamlining and opened inward whenever the guns were fired. The F-86A was outfitted with an ejection seat so that the pilot could safely leave a crippled F-86 at high speeds. The underfuselage dive brake was eliminated and the fuselage speed brakes now opened rearward instead of forward as on the XP-86s.

The F-86A-5 was the first combat-capable variant. It differed from the F-86A-1 in having a V-shaped, armored glass windscreen and heated gun compartments. It also was the first F-86 equipped with an underwing pylon capable of carrying a variety of ordnance, including 500 and 1,000 pound bombs and underwing fuel tanks of up to 206 gallons. Additionally the F-86A-5 could carry 5 inch HVAR rockets on stub launchers under the wings. Power was upgraded through the installation of the -7 engine and later the -13 engine. Both were still rated at 5,200 lbst, but they were more reliable and easier to maintain.

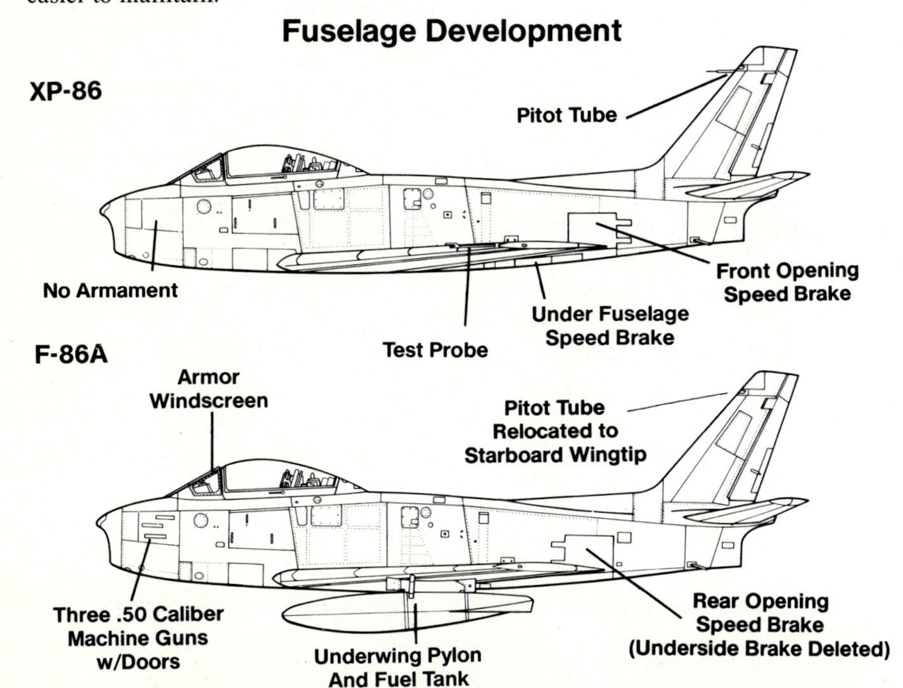

One of the external differences between the XP-86 and the F-86A was in the method by which the speed brakes operated. On the XP-86 the brakes opened frontward while on the F-86A they opened rearward. The two circular objects above the tailpipe are position lights. (George McKay)

Most of the thirty-three F-86A-1s were used for test and evaluation purposes and it was the F-86A-5 that was the first to see squadron service. The 1st Fighter Group at March AFB was the first operational unit, transitioning from Lockheed F-80s to the F-86A-5 during February of 1949. The 1st FG was charged with the air defense for the Los Angeles area, which included the North American plant where their aircraft were built. The next fighter groups to become operational with the F-86As were the 4th FG at Langley AFB which defended the nation's capitol and the 81st FG at Kirtland AFB, that covered the atom bomb plant at Alamogordo, New Mexico. In all, five USAF fighter groups were equipped with F-86As. The Air Force accepted 554 F-86As at a cost of $178,408.00 each. The first Sabres that went to reserve units were assigned to the 116th FIS/Washington ANG which transitioned to F-86As on 22 December 1950.

Of the five operational groups, it was the 4th Fighter Interceptor Group that became the most famous and it was the 4th that took the Sabre to war. When the Communist Air Forces introduced the MiG-15 into the air war over Korea in November of 1950, they immediately made all the fighter aircraft that the 5th Air Force had in the theater obsolete. The MiG could easily gain air superiority anywhere they chose to appear. The B-29 force was helpless against MiG attacks and 5th AF F-80 Shooting Stars could do little to stop the MiGs from getting the B-29s. The MiG was almost 75 mph faster than the F-80C. Had the Red generals moved the MiG force into Korea behind their successful ground offensive in the Winter of 1950, it would have made things much more interesting.

But they didn't and FEAF countered the MiG threat by asking for, and receiving the best aircraft available in the U.S. inventory — the F-86A-5 Sabre. The Air Force immediately dispatched the 4th FIG to Korea with their F-86As. After a long sea voyage aboard Navy ships, the 4th FIG put seven F-86As from the 336th FIS into operation at Kimpo Air Base (K-14), Korea on 13 December 1950. Flying an uneventful first mission on 15 December, the Sabres drew first blood on 17 December, when Baker Flight, led by LCOL Bruce Hinton, jumped a flight of MiGs near Sinuiju. LCOL Hinton shot down

The instrument panel of the F-86A was typical of jet fighters of the 1950s. This F-86A has had the gun sight removed; normally it was carried on a mount above the panel. (USAF)

Although now designated "F for Fighter," some early production F-86As came off the assembly line painted with "P for Pursuit" buzz numbers. This F-86A-1 was enroute to Ladd AFB, Alaska for cold weather tests and was painted the standard Arctic markings consisting of Red outer wing panels and tail. (NAA)

The gun compartment housed three .50 caliber machine guns and their ammunition feed system. The guns were angled to keep the ammunition from jamming as it was fed through the metal belt feed mechanisms. (NAA)

one of the MiGs for the first Sabre victory of the Korean War. It would not be the last, as Sabre pilots would rack up 791 more MiG victories before the end of the Korean Police Action in July of 1953.

There were several modifications made to the F-86A during its service life. Problems in Korea led to a replacement of the Mk 18 gyroscopic gunsight with an A-1CM sight that was linked to the AN/APG-30 radar built into the lip of the nose intake. Several model 48 F-86As that were in Korea with the 4th Fighter Interceptor Wing were converted to reconnaissance aircraft. The aircraft were flown to the Rear Echelon Maintenance Facility (REMCO) at Tachikawa, Japan for Project ASHTRAY. The ASHTRAY conversion included the removal of the gun system and ammunition supply bays and installation of a pair of K-24 cameras housed in a bulged fairing under the fuselage where the ammo bays had once been.

The cameras were mounted on a horizontal axis, which aimed them at a mirror for vertical coverage. The mirror mounting was quite crude and flimsy and many of the photos came out blurred due to camera and/or mirror vibration. But overall the project was a success. Any photo mission into or near "MiG Alley" was flown by one of the ASHTRAY RF-86As, which were painted exactly as their 4th FIW fighter cousins across the field at Kimpo. By the end of the war at least seven aircraft had been modified to RF-86A standards. Several had a "dicing" camera fitted in the lip of the nose where the gunsight radar had been mounted. These cameras were covered with small sliding doors. Some RF-86As retained two of their .50 caliber guns (with a very limited amount of ammunition) for self-defense. Although the RF-86A project was a field modification done for a combat requirement (and strictly a temporary design until RF-84Fs could be delivered) North American did build several RF-86Fs strictly for use in Korea. All the RF-86 Sabre missions in Korea were flown by 15th TRS pilots.

F-86B

The F-86B was a modification of the original F-86A with larger wheels, tires and heavier brakes which would have meant an additional seven inches added to the width of the fuselage. Air Force signed a letter of intent to buy 190 F-86Bs during 1947; however, tire and brake technology advanced to such a point that the original tire/brake size requirements were retained and the order was altered to 188 F-86As and two F-86Cs. No prototype F-86Bs were ever built.

The first operational unit to re-equip with F-86As was the 1st Fighter Group at March AFB, California which received their first Sabres in March of 1949. The rounded windscreen was only used on the F-86A-1. To maintain streamlining, the gun ports for the .50 caliber machine guns were covered by automatically opening doors. (USAF)

Late production F-86A-5s of the 93rd FIS/81st FIG on the flight line at Kirtland AFB, New Mexico. The 81st FIG was the third group to transition from F-80s to F-86s in 1949. Their mission was to defend the atomic bomb plant at Alamogordo, New Mexico. Late F-86A-5s had the gun port doors deleted.(NAA)

Nose Development

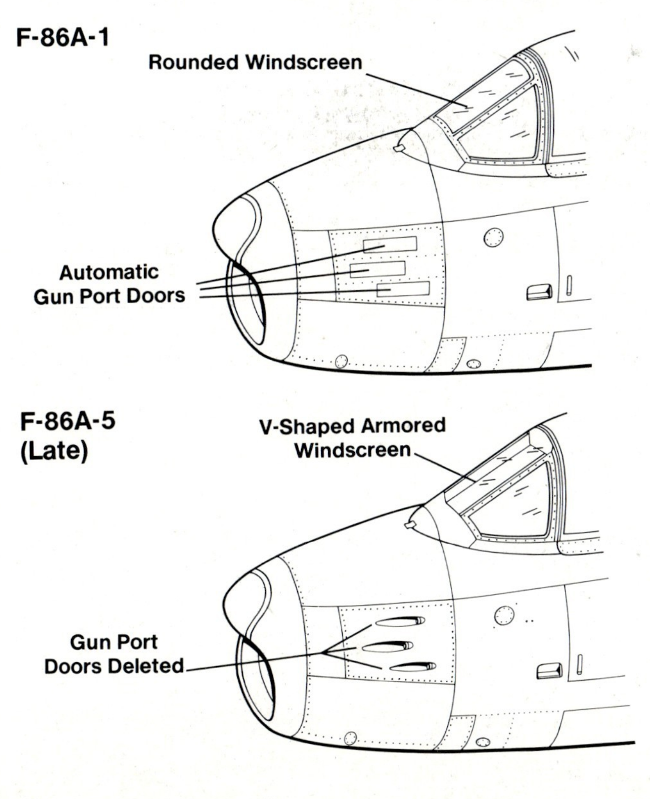

CAPT Richard D. Creighton set a World's Speed Record during a flight from San Francisco to Los Angles on 20 May 1950 with this F-86A of the 71st FIS. CAPT Creighton would later go to Korea where he became the fourth jet ace of that war. (BGEN Robin Olds)

When the Cold War heated up, the USAF transferred the F-86As of the 81st FIG to bases in England during August of 1951. The 116th FIS/81st FIG was the first squadron stationed in England since the end of World War 2. The tail colors were Yellow lightning bolts on a Blue background. The fuselage and drop tanks had Red lightning bolts with White trim. (USAF)

A colorful F-86A of the 94th FIS at March AFB during 1952. The Yellow lightning bolts and tail bands were added specifically for a special Inauguration Day flight to honor the newly elected president — Dwight David Eisenhower. (Robert L. McKinney)

This F-86A was assigned to the squadron commander of the 60th Fighter Interceptor Squadron at Westover Air Force Base, Massachusetts during 1951. The wing leading edge slats and trailing edge flaps are fully deployed. (Jim Sullivan)

A late F-86A-5 of the 23rd FIG at Presque Isle AFB, Maine. The 23rd FIG was the fifth and last USAF fighter group to be equipped with the F-86A. The aircraft was assigned to the 75th FIS and has a Blue tail and nose bands with White stars in addition to the four-color wing commanders bands on the fuselage. (David Menard)

The primary training unit for Sabre pilots was the Fighter School at Las Vegas Air Force Base (now Nellis AFB), Nevada. This F-86A carried a Black and Yellow checkered tail. The F-86A-5 could carry a pair of underwing fuel tanks or bombs, such as these 250 pound Blue practice weapons. (David Menard)

A Flight Test Center F-86A is towed on the ramp at the 1951 National Air Race, Wright-Patterson AFB, Ohio. Wright-Patterson was the home of flight test activities east of the Mississippi. The markings are Day-glo Orange with Black trim. (David Menard)

CAPT Jack Smith flew this F-86A named *TEXAS JACK*, while with the 78th FIS/81st FIW at RAF Shepards Grove during 1953. The wing and tail bands closely resemble the combat markings applied to Sabres operating in Korea. (COL Jack Smith)

An F-86A of the 75th FIS flies formation with a Grumman F9F-6 Cougar near Long Island during 1952. The Navy had also seen the advantages of the swept wing and had contracted Grumman to develop a swept wing version of the F9F Panther. The F9F-6 Cougar was the first F9F variant with swept wings. (John M. Campbell)

LCOL Glenn Eagleston flew this F-86A in Korea when he commanded the 334th FIS at Suwon in the Summer of 1951. Eagleston added two MiGs to his total of eighteen and one half German victories in the Second World War. His Sabre took a few hits in the fuselage and cockpit area from a MiG later that month, but it still brought him home. (Australian War Memorial)

The F-86A could absorb a lot of battle damage. This 4th FIG pilot is standing in what was left of his port flap after one hit from a 37MM cannon shell from a MiG-15. The Sabre still made the 200 mile flight back to Kimpo despite losing almost two thirds of the flap. (Robert L. McKinney)

Project ASHTRAY involved conversion of at least seven F-86A-5s for the reconnaissance mission by removing their guns and adding a pair of K-22 cameras in a bulged fairing under the fuselage. *Nancy* was an RF-86A serving with the 15th TRS/67th TRW at K-14 in 1952. (George McKay)

Project ASHTRAY

F-86A-5

Six .50 Caliber Machine Guns

RF-86A

One .50 Caliber Machine Gun (Some Aircraft)

Camera Bay Fairing

Camera Window

Sabre maintenance in Korea during early 1951 was usually done outdoors. These 4th FIG mechanics are doing an engine change on the Suwon ramp. The B-29 in the background was forced to land at K-13 and then turned into an open-air engine shop since there were no permanent facilities available on the field. (Irv Clark)

The surviving Project ASHTRAY RF-86As were replaced by RF-86Fs in Korea and passed to the 115th FIS/California ANG at Van Nuys Airport where they were still flying as late as June of 1959. (Brian Baker)

F-86As were phased into Air National Guard inventories beginning in 1951. These F-86As, many of which were Korean War veterans, were assigned to the 156th FIS/North Carolina ANG at Charlotte during 1953. The nose and tail flashes and wingtips were painted Red. (Jim Sullivan)

Gal-O-My Dreams was an F-86A-7 flown by LTCOL S.C. Austin when he commanded the 123rd FIS/Oregon ANG at Portland in 1955. The F-86A-7 was an F-86A that had undergone overhaul at Fresno and was modified with an AN/APG-30 radar and A-1CM gun sight. (Brian Baker)

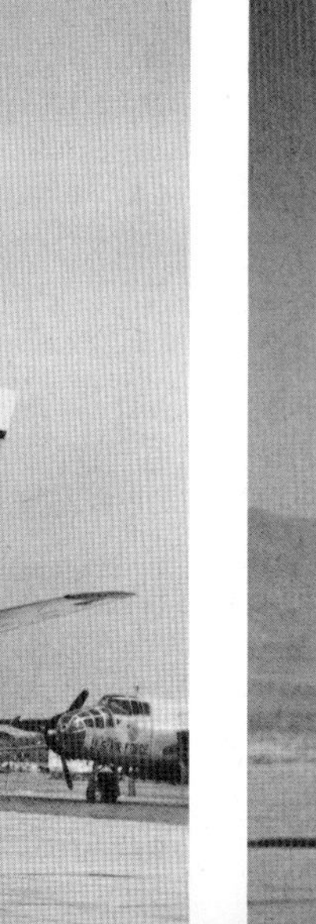

Aircraft that pulled target sleeves for "live fire" aerial gunnery practice were called target tugs and quite often were the most colorful aircraft in the Air Force. This Arizona ANG F-86A carries the standard Arizona ANG Copperheads markings plus target tug markings which included International Orange uppersurfaces. (P. Paulsen)

F-86C/YF-93A

The F-86C was designed around a requirement for a long range fighter aircraft to escort SAC bombers to Soviet targets. It retained the swept wing planform of the F-86A Sabre but had an entirely new fuselage and engine. Power for the F-86C was to have come from a 8,000 lbst Pratt & Whitney J48-P-1 engine with an afterburner. The larger J48 engine meant that the fuselage had to be enlarged both in width and length. The F-86C was forty-four feet long, some six and a half feet longer than an F-86A and much wider. The immense fuselage was fitted with additional fuel tanks that brought the total fuel capacity to 1,580 gallons.

The F-86C was armed with six 20MM cannons and had an SCR-720 search radar housed in the nose radome. With the nose now housing the radar, the F-86C had the air intakes repositioned to the sides of the fuselage using NACA-designed recessed air scoops. The fuselage also utilized the new area rule "coke bottle" shape. This pinched or "Wasp waist" design reduced drag at high Mach numbers and would be used on most of the Century Series fighters during the 1950s. The dive brakes were replaced by a large slab brake mounted under the fuselage (very similar to that found on the later F-100).

All these changes to the airframe led the Air Force to redesignate the aircraft the YF-93A. Rollout of the YF-93A came in late 1949, with the first flight taking place on 24 January 1950. The weight of the YF-93A was over 26,500 pounds in combat trim, almost twice that of the F-86A. But the 8,000 lbst J48 engine, combined with all the low-drag design features, led to a surprising top speed of 708 mph. Range on internal fuel was almost 2,000 miles, compared with a ferry range of 1,052 miles for the standard F-86A (with external fuel tanks).

Two YF-93As were built (48-317 and 48-318) and late in their service lives, both YF-93s had their air intakes redesigned with a conventional protruding scoop extending over the NACA flush intakes. The YF-93A was used to compete against the Lockheed F-90 for the role of "penetration fighter" to escort SAC bombers to Soviet targets. But development of long range jet bombers, the B-47 and B-52, with their very high speed, eliminated the need for a penetration fighter. As a result, both fighter projects were cancelled and the funds used to purchase more bombers. At some point in their service with NACA, both YF-93As had the rear fuselage modified to accept a production F-86D tailpipe and stabilizer housing. The YF-93As were used by NACA as flight test and chase aircraft well into the mid-1950s.

Late in their service lives, both YF-93As were modified with standard protruding air intakes over the NACA flush intakes. The aircraft retained the F-86A wing, tail assembly and V-shaped windscreen. (NAA)

Fuselage Development

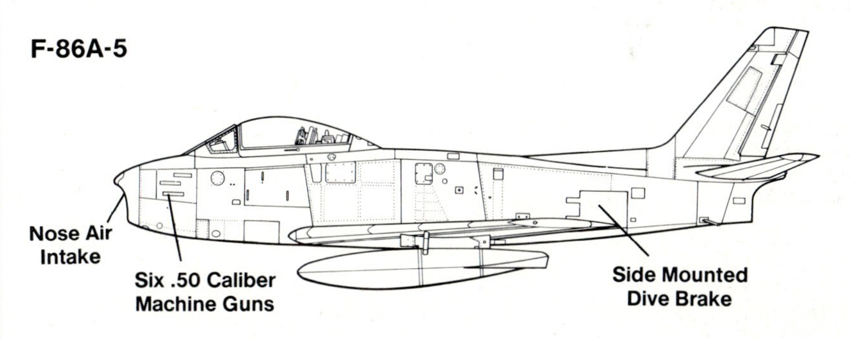

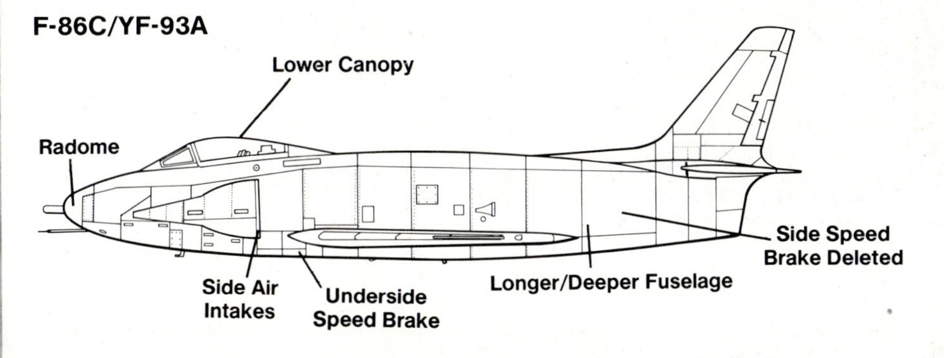

The YF-93A and F-95A share the ramp at Muroc Dry Lake Air Force Base during 1950. Both aircraft were conversions of F-86A airframes and carry service test (YF) buzz numbers on the rear fuselage. (NAA)

The F-86C/YF-93A was a North American Aviation project designed to create a long range, penetration jet fighter using existing Sabre technology. The F-86C/YF-93A used an 8,000 lbst Pratt and Whitney J48-P-1 engine with afterburner, had large side mounted NACA-designed flush air intakes and a dual wheeled main landing gear. (NAA)

Following cancellation of the penetration fighter program, both YF-93A prototypes were bailed to NACA (National Advisory Committee for Aeronautics) where they were used as chase aircraft throughout the 1950s. (Brian Baker)

F-86D

During 1949, the Air Force issued a requirement for an all-weather jet interceptor. Several manufacturers responded with formal proposals but North American's was easily the most radical. While all the other all-weather jet proposals were two seat aircraft with a pilot and radar intercept officer, the North American proposal combined the roles through the use of something radically new called a flight control computer and the concept of the "one-man interceptor" was born. While the flight control computer flew the aircraft and monitored the engine and flight controls, the pilot flew the Hughes E-3 Fire Control System to complete the intercept. When it worked correctly, it was a very good and workable system.

The computerized controls weren't the only things different about the F-86D. The F-86D was the first fighter aircraft armed solely with air-to-air rockets in place of machine guns or cannons. The twenty-four 2.75 inch Folding Fin Aircraft Rockets (FFAR) were loaded in a retractable tray in the underside of the fuselage just behind the nose landing gear well. When the pilot pulled the trigger, the tray instantly extended into the airstream and fired the correct volley of rockets.

The F-86D also differed from the earlier F-86A in the power plant. The earlier 5,200 lbst J47-GE-1 engine was replaced with a 7,500 lbst afterburning J47-GE-17 engine. All these changes left only about twenty-five percent of the original F-86A and the Air Force decided to redesignate the aircraft as the F-95A. Strangely, the F-95A/YF-86D had serials numbered after the beginning of the production run (the F-95As were numbered 50-577/578 and production began with the F-86D-1 50-455). Although some 2,500 pounds heavier than the F-86A, the F-86D was actually faster and the F-86D broke the World Speed Record twice within a year. CAPT Slade Nash broke the previous record when he went 698.5 mph in November of 1952. This record was short-lived and LCOL William Barnes broke the record in July of 1953 with a speed of 715.6 mph.

Use of the Hughes E-3 Fire Control System, AN/APG-36 search radar with its thirty inch radar dish and afterburning J47-GE-17 engine, necessitated the use of a completely redesigned fuselage. The new fuselage differed from the F-86A in that it was some three feet longer and was somewhat deeper and wider. The air intake was moved under the radome and was much wider than the F-86A intake. Although the first prototype had the standard F-86A sliding canopy and windscreen, production F-86Ds used a clamshell opening canopy. The F-95A/YF-86D retained the wing with leading edge slats, landing gear, dive brakes and basic flight controls of the F-86A.

One major change in the flight controls was the use of an all-flying tail unit. The horizontal stabilizer and elevators were combined in a single "slab" stabilizer and this entire "slab" moved, giving much greater elevator control. It meant that the pilot retained normal control of the aircraft in the transonic/supersonic speed range instead of suffering control reversal, common to aircraft having a standard stabilizer/elevator control system.

Throughout its service life, the F-86D was constantly upgraded, especially in its electronics. The F-86D-5 introduced the Hughes E-4 Fire Control System. The D-25 had provision for drop tanks. The D-35 had an omni-directional ranging radar. The D-45 introduced a drag parachute to aid in stopping the F-86D on short runways. The D-60 was powered by a 7,950 lbst J47-GE-33 engine. The final upgrade was the most radical and resulted in a new designation — the F-86L.

Darleen Craig, an employee at the NAA plant in Inglewood, poses with the F-95A during the rollout ceremony in September of 1949. The aircraft was later redesignated as the YF-86D. The prototype retained the sliding canopy and V-shaped windscreen of the F-86A. (NAA)

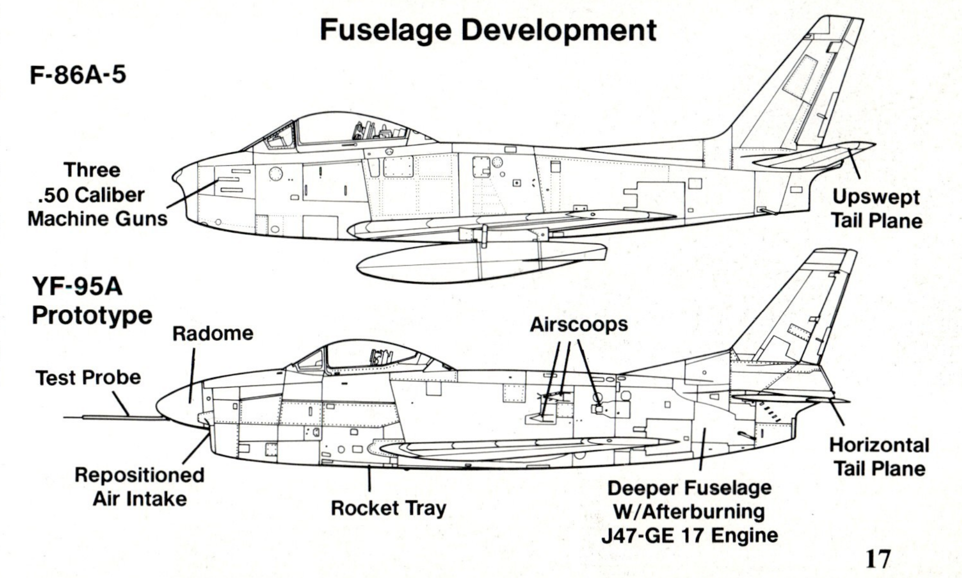

F-86L

The F-86L was not a new production aircraft but rather a rebuild of existing F-86D airframes modified with several totally new systems. The wing was replaced with a longer-span, slatted wing (then in use on the F-86F-40), all the electronics were brought up to the latest "state of the art" technology, including stripping the entire aircraft of its existing wiring and replacing it with all new wiring. A new electronic data-link between the air defense ground controller and the pilot was added to the aircraft. It was known as the SAGE system (Semi-Automatic Ground Environment). The SAGE equipment allowed the interceptor pilot instant access to any and all information available from the ground controller. SAGE equipment included an AN/ARR-39 data link receiver, AN/ARC-34 command radio, AN/APX-25 identification radar and a new glide slope receiver. The SAGE antenna under the forward fuselage was the main external difference between the F-86D and F-86L.

The SAGE equipment was tied into the E-4 Fire Control System and automatically computed the proper lead-collision attack course. These modifications were conducted under Project FOLLOW ON at the North American facilities at Inglewood and Fresno, California. The first flight of the F-86L took place in late Spring of 1956, with the 49th FIS taking delivery of the first operational F-86L in October. The Inglewood plant modified 575 late production F-86Ds (the last F-86D came off the assembly line in September of 1955) to F-86L standards, with the Fresno plant doing an additional 452 aircraft. The Air Force accepted a total of 827 F-86Ls by the time the program ended.

The F-86D quickly became known in the Air force as the "Sabre Dog." The F-86Ds flew air defense missions throughout the world with USAF units active in Europe and the Far East. Serving alongside Lockheed F-94 Starfires and Northrop F-89 Scorpions, it was the "one-man interceptor" from North American that was clearly the first choice for the Air Force. Of the thirty squadrons that comprised the Air Defense Command during the mid-1950s, twenty were equipped with F-86Ds!

F-86Ds and F-86Ls were also put into the Military Assistance Program and served with a number of friendly foreign air forces. Denmark purchased sixty F-86Ds between 1958 and 1960 assigning them to Nos 723 and 726 *Eskadrillerne*. Denmark was the only NATO air force to be equipped with F-86Ds. Four SEATO air forces were equipped with F-86Ds — South Korea, Japan, Nationalist China, and The Philippines. The Royal Thai Air Force purchased F-86Ls as they became available. The USAF Air National Guard equipped twenty-six air defense squadrons with F-86D or L aircraft. The last F-86D/L mission was flown by the 196th FIS/California Air National Guard during mid-1965, just before they transitioned into supersonic Convair F-102As.

The second YF-86D prototype in flight over Muroc Dry Lake AFB during 1950. The extremely large size of the tailpipe area was caused by the use of a computerized, afterburning variant of the J47 engine. In service, the F-86D series was commonly called the "one man interceptor." (NAA)

This F-86D on the snow covered ramp at Ladd AFB, Alaska during 1953 carries full Arctic markings in Red. All new aircraft and systems purchased by the USAF were sent to either Ladd AFB or to the climatic hanger at Eglin AFB, Florida for cold weather testing. (NAA)

Canopy Development

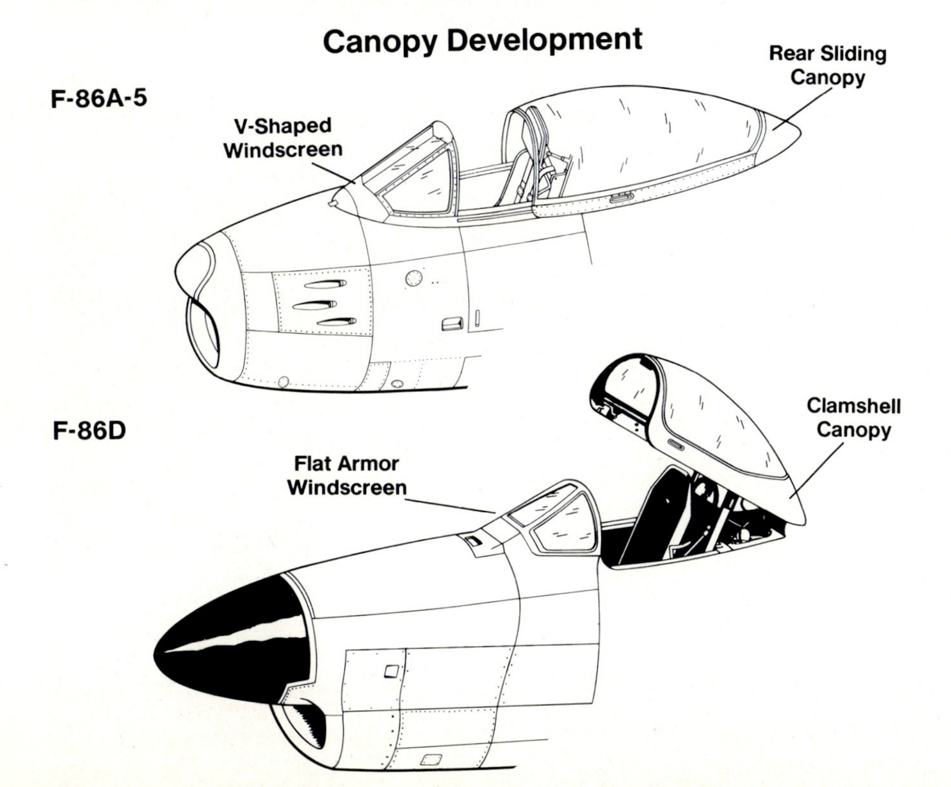

An F-86D in the engine test shed at the NAA facility in Inglewood. A complete test of all systems, including the engine, was required before an aircraft was accepted by the Air Force. The sliding doors on the shed were shaped to fit the F-86D fuselage, sealing in most of the engine noise. (Fernando Silva)

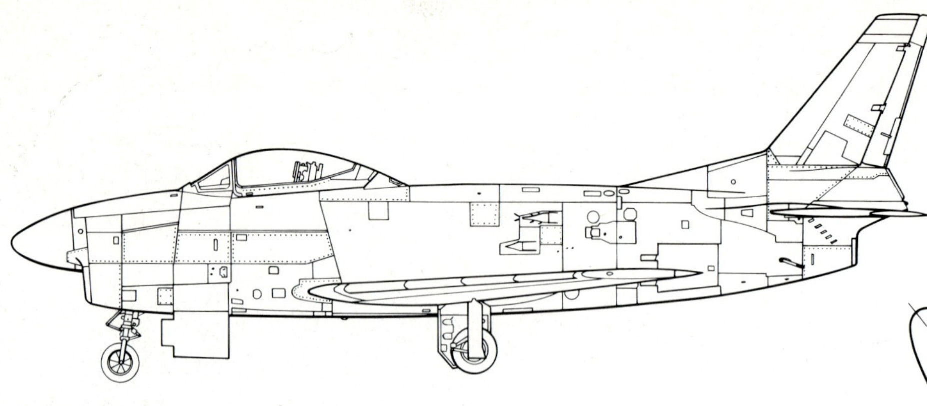

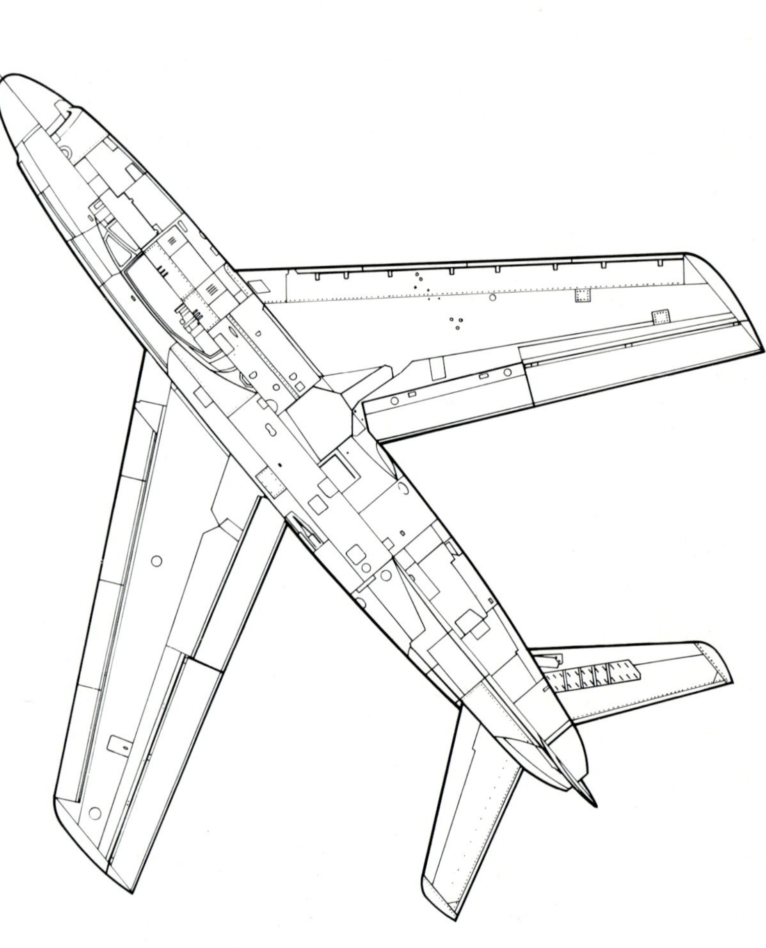

Specifications

North American Aviation F-86D-45 Sabre

Wingspan . . . . 37.54 feet
Length . . . . 37.12 feet
Height . . . . 15 feet
Empty Weight . . . . 13,498 pounds
Maximum Weight . . . . 18,160 pounds
Powerplant . . . . One 7,650 lbst General Electric J47-GE-33 turbojet engine.

Armament . . . . Twenty-four 2.75 inch FFAR rockets.

Performance
Maximum Speed . . . . 679 mph
Service ceiling . . . . 49,600 feet
Range . . . . 1,022 miles
Crew . . . . One

The North American ejection seat used in the F-86D series was different from the type of seat used in the F-86 day fighter variants. (NAA)

North American mechanics load the rocket tray on this overall Day-glo Red F-86D-1 Sabre at Edwards Air Force Base in the early 1950s. The fairings on the wing and fuselage house cameras to record the results of air-to-air rocket firing tests. (NAA)

The only official aerobatic team in the Air Defense Command during the mid-1950s was the Sabre Knights from the 325th FIS at Hamilton AFB, California. The team flew their F-86Ds at air shows throughout the nation, including the Detroit Airport air show in July of 1955. (David Menard)

An F-86D-35 of the 93rd FIS at Kirtland AFB, New Mexico in 1957. The D-35 replaced the older radar with an omni-directional radar set. The lightning bolts on the fuselage and the wingtips are in Red. (USAF)

F-86Ds move into the radar installation point on the F-86D assembly line at Inglewood. The use of a 30 inch radar dish antenna meant that the intake had to be widened horizontally to provide sufficient air to feed the afterburning J47-GE-17 engine. (NAA)

A pilot and crew chief scramble to their 16th FIS F-86D during an alert at Taipai Air Base, Formosa. Far East Air Force F-86Ds stood alert on Formosan bases alongside day fighter units as part of the Composite Strike Force during the Formosa Crisis of 1957. (USAF)

An F-86D of the 525th FIS/86th FIW based at Bitburg AB, West Germany during June of 1957. F-86Ds were assigned the air defense mission throughout the Free World in the mid-1950s, serving as the main air defense weapon against Soviet bomber attacks. The nose and tail are Dark Blue and White. (R. Anderson)

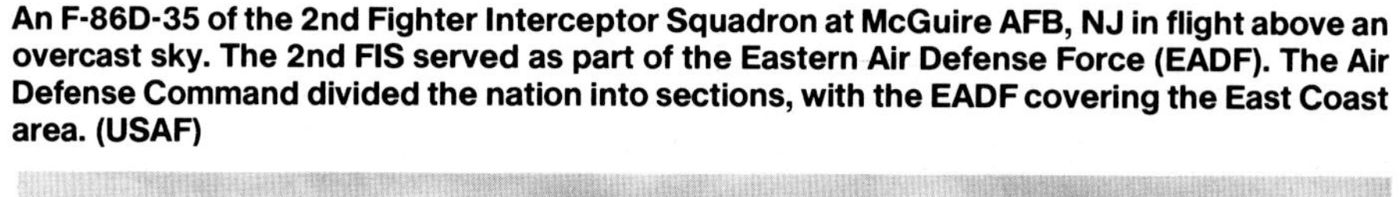

An F-86D-35 of the 2nd Fighter Interceptor Squadron at McGuire AFB, NJ in flight above an overcast sky. The 2nd FIS served as part of the Eastern Air Defense Force (EADF). The Air Defense Command divided the nation into sections, with the EADF covering the East Coast area. (USAF)

F-86Ds of the 357th FIS on the ramp at Nouasseur Air Base, French Morocco during early 1960. Following a seven year deployment to Morocco, the 357th went home to George AFB, California during July 1960. (Major R.G. Smith)

Tail Development

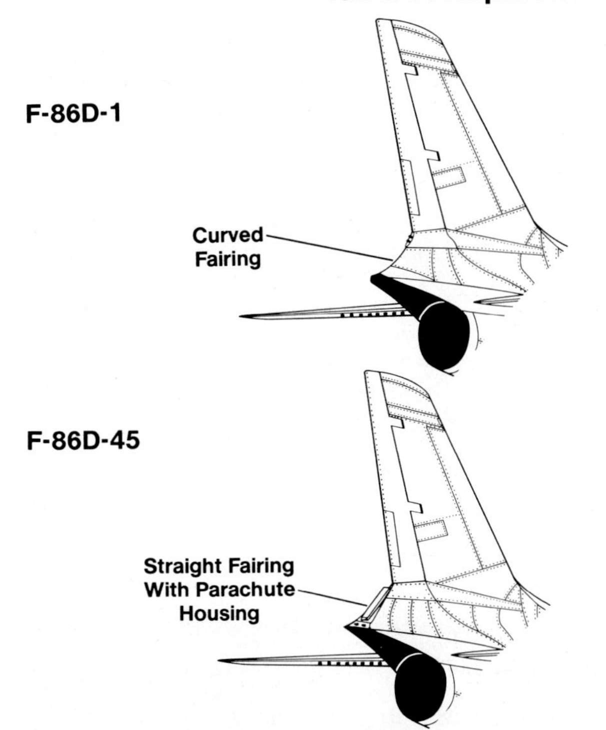

The last F-86D-45 off the assembly line went to the 456th FIS at Castle AFB, California. The D-45 introduced the drag parachute to the F-86D series which was housed in a fairing at the base of the rudder. (BGEN K.W. Bell)

The pilot of this F-86D-50 of the 86th Fighter Interceptor Squadron has the instrument hood pulled closed for a simulated night rocket attack against the target sleeve at the 1956 Yuma Rocket Meet. (MGEN H.E. Collins)

F-86D-50s of the 431st FIS Red Devils carried highly colorful markings. This Sabre Dog is on the ramp at Wheelus AFB, Libya during 1956. The 431st was one of two units flying F-86Ds with SAC from Spanish bases during 1958-59. (Bob Esposito)

F-86Ds of the 20th Air Division line the ramp at the 1956 Yuma World Wide Rocketry Meet held at Yuma AFB, Arizona. The aircraft in the foreground is a squadron commander's aircraft with commander's bands on the rear fuselage. Today this meet is known as the William Tell Missile Meet. (NAA).

An F-86D-40 of the 497th FIS at Geiger Field, Washington during 1955. Colorful markings were normal for ADC Sabre Dog units. The nose and tail flash are Black and White and the serial was carried in a nonstandard location, so as not to interfere with the tail markings. (Peter Bowers)

An F-86D of the 5th FIS on the ramp at McGuire Air Force Base, New Jersey. The flying A on the drop tank identified this aircraft as being part of A flight. The 2.75 inch FFAR rocket tray is in the lowered or firing position. (LCOL E. Bosetti)

This F-86D-60 was flown by the commander of the 539th FIS at Stewart AFB, New York. The F-86D-60 featured improved radio equipment and other avionics and was the final production variant of the F-86D. (E. Sommerich)

An F-86D of the 31st FIS fires a volley of 2.75 inch rockets during the 1956 Yuma Rocket Meet. For the meet, cameras were installed under each wing, outboard of the underwing fuel tanks. (Robert Wainwright)

"SWEET MUDDER" was an F-86D-25 flown by the commander of the 329th FIS based at George AFB, California. Their delta winged unit insignia was a sign of things to come since the squadron transitioned from the F-86D to the Convair F-102A during 1958. (Peter Bowers)

A F-86D of the Air Training Command at Yuma AFB during 1956. During 1956, the first F-86Ds were pulled from active service to begin Project FOLLOW-ON, the conversion of F-86Ds to the F-86L standard. (Brian Baker)

The Danish Air Force operated a number of F-86D-30s during the 1960s assigning them to No 726 *Eskadrilleskjold*. The nose and tail markings were Medium Blue. The aircraft were modified to carry AIM-9 Sidewinder AAMs after they were delivered. (Merle Olmsted)

A polished Natural Metal F-86D of the 173rd FIS/Nebraska Air National Guard flies over the countryside near Lincoln. The aircraft carried the ANG logo in Black on the fuselage, fin, and upper starboard, lower port wings. (National Guard Bureau)

The Republic of Korea (ROK) Air Force received a number of F-86D-50s modified with AIM-9 Sidewinder air-to-air missile rails under the inboard wing panels. The small antenna on top of the nose just behind the radome is a TACAN antenna. (ROKAF)

This well maintained F-86D-35 was assigned to the 185th FIS, Oklahoma Air National Guard stationed at Will Rogers Field, Oklahoma City. After a complete overhaul, former ADC F-86Ds were assigned to ANG units beginning in April of 1958. (Richard Lindsey)

This F-86D was flown by LTCOL William Fairbrother when he commanded the 324th FIS at Sidi Slimane, Morocco in 1958.

This Blue tailed F-86F was flown by COL Chesley Petersen, commander of the 48th FBW at Chaumont, France during 1953.

An F-86F-40 flown by the Republic of Korea Air Force aerobatic team based at Suwon, Korea during 1961.

A Canadair CL-13B Sabre Mk 6 of No 434 Squadron, Royal Canadian Air Force at Cold Lake in 1959.

A Day-Glo Orange F-86H-5 of the 1st FDS/413th FDW at George Air Force Base, California during 1956.

An F-86A-7 target tug assigned to the Utah Air National Guard.

MAJ Ed Heller flew *HELL-ER BUST X* when he commanded the 16th FIS at Suwon, Korea during January of 1953.

(Starboard Side)

Temptation was an F-86E flown by LT Donald McLean with the 334th FIS at Kimpo, Korea during 1952.

A Canadair Sabre F.1 of No 234 Squadron, Royal Air Force at Oldenburg, Germany in 1954.

"TWEETY" was an F-86D Sabre Dog of the 25th FIS at Naha, Okinawa during 1955.

An F-86L of the 83rd FIS on the ramp at Hamilton AFB during 1958. The F-86L was a rebuilt late production F-86D with the SAGE system installed. The SAGE antenna is visible under the nose, near the rocket tray. (Jim Sullivan)

This F-86L was assigned to the 456th FIS at Castle AFB during 1958. Under project FOLLOW-ON, NAA rebuilt a total of 827 F-86Ds to the F-86L standard. The first flight of an F-86L took place in January of 1957 and the last was delivered in November of that same year. (BGEN K.H. Bell)

F-86L SAGE Antenna

F-86D

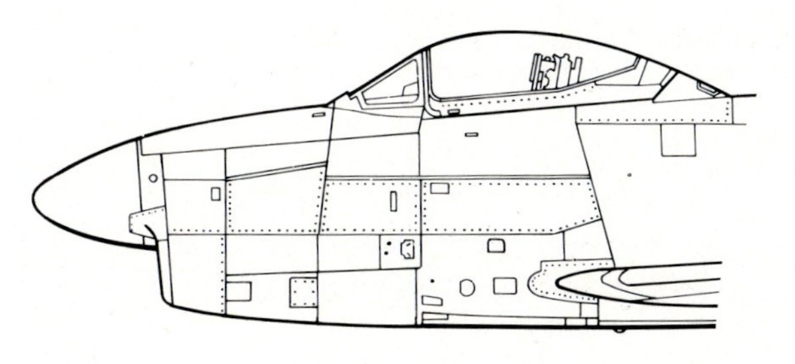

F-86L

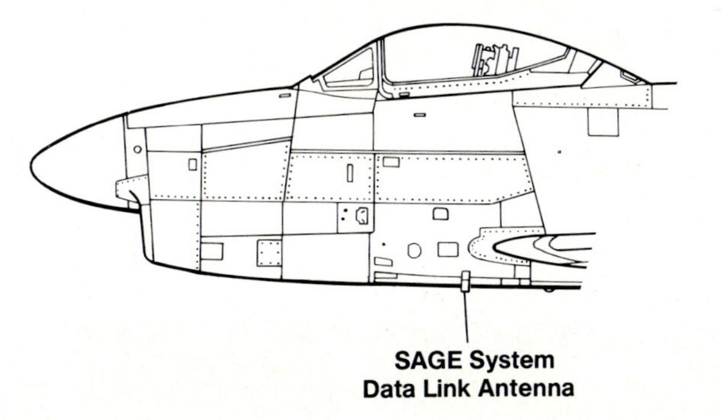

Only one foreign nation flew the F-86L. The Royal Thai Air Force operated a number of F-86Ls with at least one squadron still active at Bangkok during the early years of the Vietnam War. (David Menard)

A trio of F-86Ls of the 157th FIS/South Carolina Air National Guard based at McEntire ANG Base during 1959. The F-86D block number changed one digit when the aircraft was converted to L standards. An F-86D-50 became an F-86L-51. (Jim Sullivan)

Most F-86Ls which served with Air Guard units were assigned the air defense role. This F-86L-56 was assigned to the 194th FIS/144th FIW, California Air National Guard at Fresno during 1962. The nose, tail and wing bands are Day-glo Orange. (Robert F. Dorr)

F-86E

Initially the F-86E was simply an F-86A fitted with an all-flying tail in place of the standard stabilizer/elevator unit. The F-86E-1 retained the same wing, engine and V-shaped windscreen as used on the F-86A-5. The F-86E-6 was a USAF designation for sixty Canadair Sabre Mk 4s that were purchased during 1952 to augment the depleted Sabre forces operating in Korea. The F-86E-10 introduced an optically ground, armored, flat windscreen in place of the earlier V-shaped windscreen. Finally the F-86E-15 was an F-86F airframe with a J47-GE-13 engine.

Although most of the F-86Es produced saw service with active Air Force units, including both the 4th and 51st FIGs in Korea, they were rapidly replaced once the next Sabre variant, the F-86F, became available and the F-86Es were phased into service with Air National Guard units. On 1 February 1956, the 119th FIS/New Jersey ANG transitioned to the F-86E. The Michigan ANG was the recipient of many of the Canadair-built F-86E-6s, most of which were Korean combat veterans. North American built a total of 456 F-86Es at a flyaway cost of $219,457.00 each.

Canadair Ltd. of Canada was another major manufacturer that built the Sabre. Canadair signed a license agreement with North American Aviation in August of 1949 to build the Sabre in Canada. The aircraft would be an exact copy of the latest F-86 design at the time production began. In addition, any USAF improvements in the aircraft would also be incorporated in the Canadair program. The aircraft were to be built at the Canadair facility in Cartierville, near Montreal. Under the agreement, North American Aviation would provide all the necessary drawings and parts to construct the jigs and tools needed to manufacture the Sabre. The first parts were sent to Canadair in 1949 and assembly of the first Canadair CL-13 Sabre (19101) was completed in August of 1950. It was a Canadian-built F-86A-5 and flew for the first time on 9 August 1950. Several days later Canadair Chief Test Pilot Al Lilly became the first Canadian to break the speed of sound when he put the one and only Sabre Mk 1 through the sound barrier.

Full Canadair production began with the Sabre Mk 2 (only one Mk 1 was ever built). With the exception of RCAF required equipment such as radios, the Mk 2 was identical to the North American-built F-86E-1 with the "all-flying tail" and was powered by a 5,200 lbst J47-GE-13 engine. Canadair built a total of 350 Mk 2 Sabres and it was from this initial production batch that the USAF purchased sixty aircraft for use in Korea. These were built during early 1952, then flown to Fresno where USAF required equipment was installed. Once this was competed, the aircraft were designated as F-86E-6s.

The next production variant was powered by a 6,000 lbst Avro Orenda 3 engine under the designation Sabre Mk 3. Problems with the Orenda 3 engine led to program delays and only one Mk 3 was built. Production of the Canadair Sabre continued with the Sabre Mk 4, a license copy of the NAA F-86E-10. Canadair built some 438 Sabre Mk 4s, equipping both RCAF and RAF units (aircraft in RAF service were designated Sabre F.1s).

The F-86E prototype was painted in a Red and White scheme for use as a North American chase plane during 1955. The F-86E differed from the earlier F-86A in having an "all-flying" tail with a one piece slab stabilizer instead of a stabilizer/elevator combination. (NAA)

Tail Development

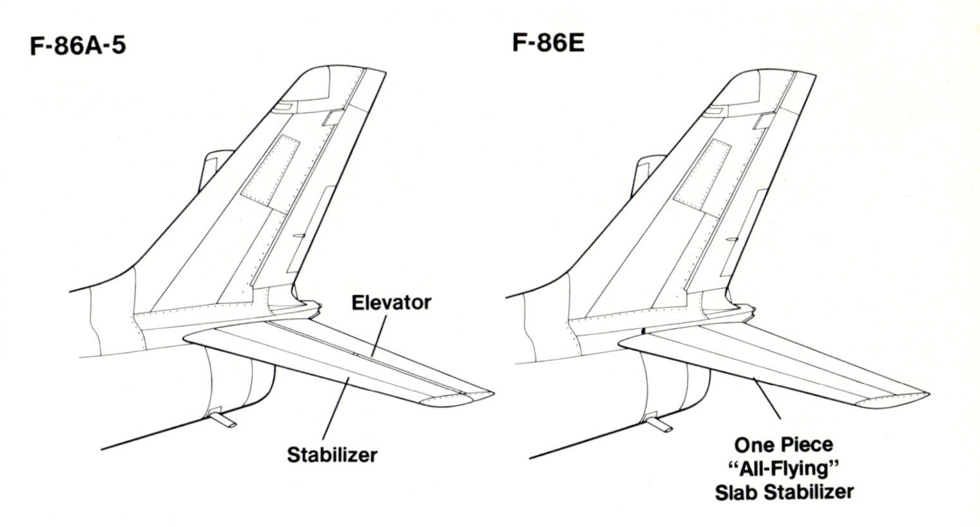

The first combat unit to re-equip with the F-86E was the 51st FIG at Suwon (K-13). The F-86E equipped the 16th and 25th Fighter Interceptor Squadrons during September of 1951. The Black outlined Yellow fuselage bands were a 51st FIG marking until adopted by FEAF for all F-86s in 1952. (NAA)

COL Fred Ascani flew this F-86E to a new World Speed Record during 1951. The aircraft went straight from speed test to combat when it was transferred to the 51st Fighter Interceptor Group in Korea. (Norm Taylor)

CAPT James Horowitz climbs aboard his F-86E named *Slow Boat To China* at Kimpo, the home base of the 4th FIW. He would later write a book about the war that was made into a popular movie titled "The Hunters." The aircraft has a V-shaped windscreen, identifying it as an F-86E-1 or E-5. (Karl Dittmers)

An F-86E-10 of the 42nd Fighter Interceptor Squadron based at Orchard Place AFB, Chicago (later O'Hare International) during 1953. The F-86E-10 was the first Sabre variant to use the optically flat armored windscreen. (Peter Bowers)

Pilots man their F-86Es of the 25th FIS on the flight line of Suwon Air Base, Korea during the Spring of 1953. The aircraft in the foreground is an F-86E-10 with an optically flat armored windscreen. (COL Hank Buttlemann)

DOLPH'S DEVIL was an F-86E-1 flown by CAPT Dolph Overton while assigned to the 16th FIS in the Fall of 1952. CAPT Overton became the twenty-fourth ace of the Korean War when he shot down two MiG-15s near Antung on 21 January 1953. (Dolph D. Overton III)

A new production F-86E-5 assigned to the 4th FIG at K-14 during the Fall of 1951. As they were received, F-86Es were mixed in with the F-86As already in service. For combat flights, however, the aircraft were not mixed since the F-86E was much faster than the older F-86A. (NAA)

Windscreen Development

F-86E-1/5

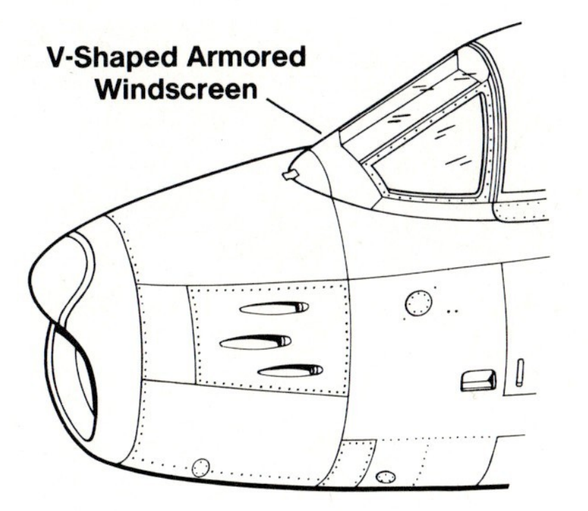

F-86E-10

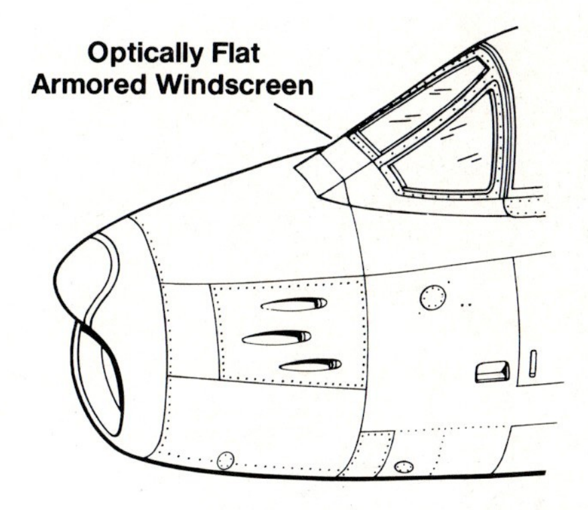

Mr. Bob Hoover confers with the North American Tech Rep alongside an F-86E of the 25th FIS on the K-13 flight line. Hoover was in Korea to demonstrate what the F-86E could do. The second aircraft in line, *THIS'LL KILL YA*, was the World Speed Record holder (51-2721). (NAA)

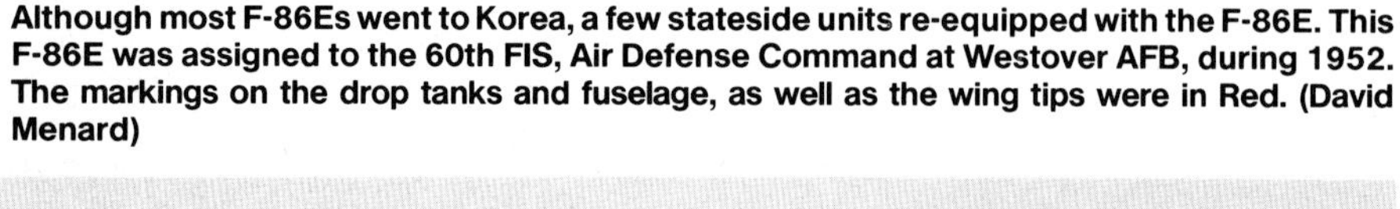
Although most F-86Es went to Korea, a few stateside units re-equipped with the F-86E. This F-86E was assigned to the 60th FIS, Air Defense Command at Westover AFB, during 1952. The markings on the drop tanks and fuselage, as well as the wing tips were in Red. (David Menard)

This F-86E is undergoing an engine change. The rear fuselage has been pulled away at the "break point" and placed on a special cart. The Sabre was assigned to LT Henry Buttlemann who became the youngest ace of the war when he shot down his fifth MiG on 30 June 1953. (COL Henry Buttlemann)

An F-86E-15 of the 121st FIS/ Washington D.C. Air National Guard. The F-86E-15 was an F-86F fuselage with an F-86E-13 engine and '6-3' fenced wings. The nose and tail bands are Yellow with Black trim and White stars. (Peter Bowers)

An F-86E-1 of the 170th FIS, Illinois Air National Guard. As the F-86E was phased out of active service many were passed to the Air National Guard. The 170th converted from F-51D Mustangs to F-86E Sabres during 1953. The unit markings are in Yellow. (Collect Aire Photo)

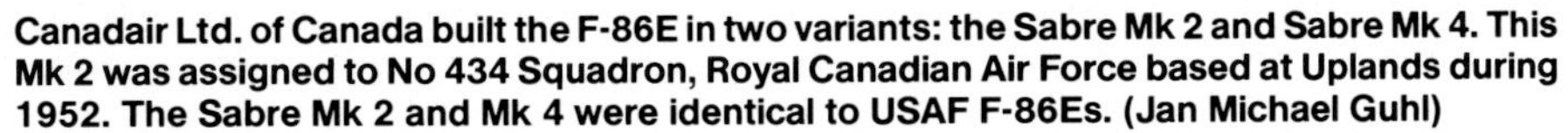

Canadair Ltd. of Canada built the F-86E in two variants: the Sabre Mk 2 and Sabre Mk 4. This Mk 2 was assigned to No 434 Squadron, Royal Canadian Air Force based at Uplands during 1952. The Sabre Mk 2 and Mk 4 were identical to USAF F-86Es. (Jan Michael Guhl)

This Canadair Sabre Mk 4 was enroute to the Italian Air Force during 1958. A number of ex-Royal Air Force Mk 4s were overhauled by Airwork Inc., painted with NATO camouflage and temporary USAF markings, then ferried to Italy. (Merle Olmsted)

This Sabre Mk 4 was used by the Italian Air Force aerobatic team, *Cavallino Rampante*. The fuselage and wings were White, the nose flash was Red and the tail was Blue with White stars. The undersurfaces were painted Red/White/Green. (Jan Michael Guhl)

F-86F

The F-86F was the ultimate clear air, day fighter variant of the Sabre. It was built in four basic sub-types — narrow chord wing with Leading Edge Slats (LES), "6-3 wing" with fences, with atomic capabilities, and with extended span wings with LES. One of the biggest gripes that combat pilots in Korea had about the Sabre was its lack of engine power. It was a great flying aircraft and very safe with its armored cockpit and backup systems. But all these safety additions meant added weight and added weight meant a decided disadvantage with the MiG in rate of climb and service ceiling. As a result, General Electric began a development program with the non-afterburning J47 which resulted in the 5,910 lbst J47-GE-27 engine. The increase in power raised the rate of climb to 9,300 ft/min. Although still less than the MiG-15, it did make the Sabre pilots more competitive against the Soviet fighter. Top speed also increased with a sea level speed of 693 mph. With the F-86F-10, a new A-4 radar ranging gunsight was added to the fire control system.

The F-86F-25 variant is regarded by many as the ultimate day fighter of the early 1950s. Initially the F-86F-25 and -30 were simple modifications with an additional hard point added to each wing for carriage of a second drop tank or 1,000 pound bombs. But in August of 1952, North American Aviation engineers tested a new wing design. The wing chord was extended six inches at the root and three inches at the wingtip. The extension was done on the wing leading edge and, at the same time, the leading edge slats were removed and replaced with a fixed leading edge. A small wing air flow fence was also added to the wing (approximately seventy percent of the span). This fence helped guide airflow over the wing and reduced transonic buffet. The result — a tighter turning radius at high Mach numbers.

Sabre pilots now had an aircraft that could easily out dive the MiG, could now turn with and inside the MiG and had almost as great a rate of climb. The MiGs still had an altitude advantage, which saved a lot of them from destruction in Korea. With all the other advantages the Sabre pilot had (self-sealing fuel tanks, armored cockpit, radar gun sights and much better training) — the result in combat was predictable. In May and June of 1953, Sabre pilots, most of them flying new F-86Fs, shot down a total of 133 MiGs for the loss of one Sabre. Most of the credit has to go to the Sabre pilots who took full advantage of all the new modifications to the aircraft.

The final two major day-fighter sub-variants were the F-86F-35 and F-86F-40. The F-35 featured the Low Altitude Bombing System (LABS) which was capable of "loft-bombing" a small atomic weapon. The F-86F-40 was a development for the Japanese Air Self Defense Force which featured the 6-3 wing. The 6-3 wing was fitted with leading edge slats and a one foot extension at the wingtip. The longer wing with slats made for a very stable aircraft at low speeds, which was needed for all the Military Assistance Program pilots from nations all over the world that were going to be equipped with Sabres of various types. This F-40 wing was also adopted by the USAF for the Sabres used at the Nellis AFB Fighter School for training both U.S. and foreign pilots.

At the peak of their service, the USAF had sixteen wings equipped with the F-86F. North American Aviation built a total of 1,959 F-86Fs at a fly-away cost of $211,111.00 each. In addition to the North American total, Mitsubishi built another 300 F-86F-40s for the Japanese Air Self Defense Force. Strangely, the F-86F was not phased into Air National Guard service in great numbers. Most F-86Fs going into the MDAP inventory were sold to other nations.

TF-86F

To produce a two seat training variant of the Sabre, North American took a pair of F-86F-30s off the assembly line for conversion to two-seat TF-86F "transonic trainers." The fuselage was extended almost sixty-three inches and a second cockpit with full flight controls was added. The wing was also moved forward eight inches for increased stability and a very large clamshell-type canopy was fitted. The first TF-86F (52-5016) was finished in early December of 1953 and made its first flight on 14 December. Tests indicated that performance was nearly equal to that of a standard fighter F-86F.

The tests were cut short when the first TF-86F crashed on 17 March 1954 and the Air Force immediately authorized a second prototype. The second TF-86F (53-1228) was completed in August of 1954 and made its first flight on 5 August. The second TF-86F differed from the first in that it had underwing hard points for drop tanks or bombs. It also had a pair of .50 caliber machine guns in the nose for gunnery practice. The second TF-86F went through extensive testing before being delivered to the Fighter School at Nellis on 31 January 1955. In any event, it was never used in the training mission that it was developed for, mainly due to development of the TF-100 trainer. The aircraft was used at Edwards AFB Flight Test Center in the chase aircraft role well into the 1960s.

RF-86F

In Korea, reconnaissance missions were flown by a variety of aircraft types, from T-6s to RB-45Cs. Any mission flown into MiG Alley (or beyond) had to be flown by a fighter type aircraft to have any chance of survival. It did not take long for MiG pilots to realize

MAJ C.L. Hewitt accepts the first F-86F (51-2850) for the Air Force from J.S. Smithson, Vice President of North American. The F-86F differed from the F-86E in the power plant. The F-86F used the 5,910 lbst J47-GE-27 engine. (USAF)

that a lone F-80 near the Yalu River was probably an unarmed RF-80 reconnaissance aircraft — an easy target for the MiG-15! FEAF and the crews with the 67th Tactical Recon Wing at Kimpo came up with the idea of modifying a Sabre to carry cameras for missions into MiG Alley.

Project ASHTRAY saw several older F-86As from the 4th FIW pulled from combat and sent to Tachikawa, Japan where FEAF mechanics removed their armament and installed a pair of K-24 cameras. Although the ASHTRAY conversions worked, improvements were needed. North American, under Project HAYMAKER, built several RF-86F conversions which were much more stable camera platforms. Some of these aircraft also retained a minimum defensive armament of two .50 caliber machine guns. Although only a handful saw service with USAF units, both the Japanese and South Korean air forces had squadrons of RF-86Fs in service.

CL-13A Sabre Mk 5 and CL-13B Sabre Mk 6

Canadair production of the F-86F series began with the CL-13A Sabre Mk 5. The Canadair CL-13A was not just a licensed copy of the F-86F since it was powered by the 6,355 lbst Avro Orenda 10 engine and incorporated the '6-3' wing with wing fences. The first Sabre Mk 5 was completed on 21 July 1953 and flown on 30 July. Canadair built a total of 370 Mk 5s, seventy-five of which were delivered to the Federal German *Bundesluftwaffe*. Many Canadair Sabres of all Mks were modified with British Martin Baker ejection seats in place of the standard North American seat.

The final Canadair-built variant was the CL-13B Sabre Mk 6. The Mk 6 was powered by the 7,275 lbst Orenda 14 engine making the Mk 6 the fastest Sabre variant, with a top speed in excess of 710 mph at sea level, a full 40 mph faster than the F-86A. Rate of climb was 11,800 ft/min, some 1,700 ft/min greater than the MiG-15. Initial production of the Mk 6 incorporated the '6-3' wing with fences, but the poor stall characteristics of the wing led Canadair engineers to replace it with a wide chord '6-3' wing with LES. Canadair production of the Sabre continued until 9 October 1958 when the last Sabre Mk 6 for West Germany came off the assembly line at Cartierville. It was the last of 655 Mk 6s and the 1,815th Canadair Sabre built.

F-86Fs were rushed to Korea during July of 1952, including this F-86F-1 of the 336th FIS. The aircraft was fitted with the extended chord hard edged wing and wing fences. The F-86F easily out performed the earlier F-86s. (USAF)

The Canadair equivalent to the F-86F was the Sabre Mk 5, although the Canadian built aircraft were powered by an Orenda 10 engine in place of the General Electric J47. This Sabre Mk 5 was assigned to No 416 Squadron, RCAF based at Grostenquin. (Brian Baker)

Wing Development

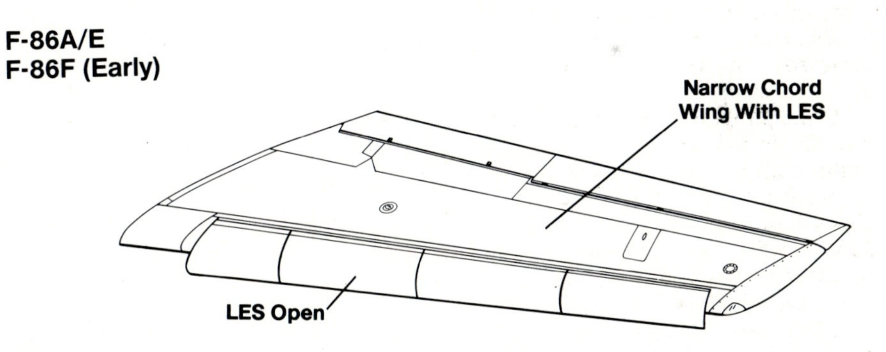

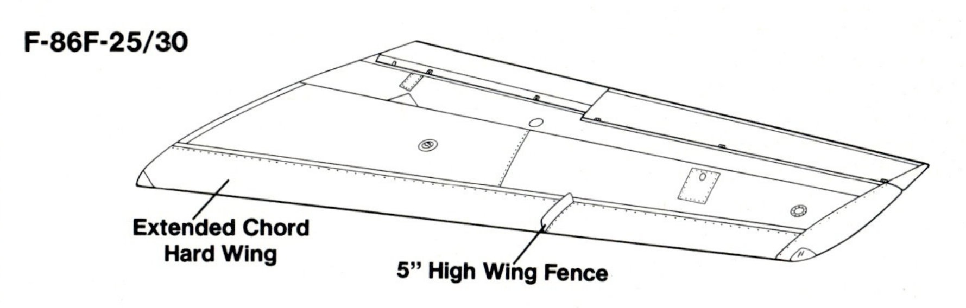

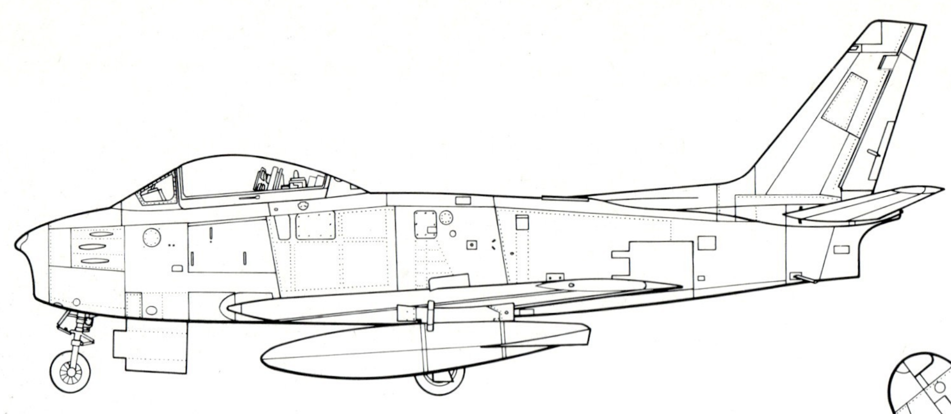

Specifications

North American Aviation F-86F-25 Sabre

Wingspan .... 37.54 feet
Length .... 37.12 feet
Height .... 14.79 feet
Empty Weight .... 10,950 pounds
Maximum Weight .... 20,650 pounds
Powerplants .... One 5,910 lbst General Electric J47-GE-27 turbojet engine

Armament .... Six .50 caliber machine guns and up to 2,000 pounds of underwing stores.

Performance
Maximum Speed .... 688 mph
Service ceiling .... 48,000 feet
Range .... 1,317 miles
Crew .... One

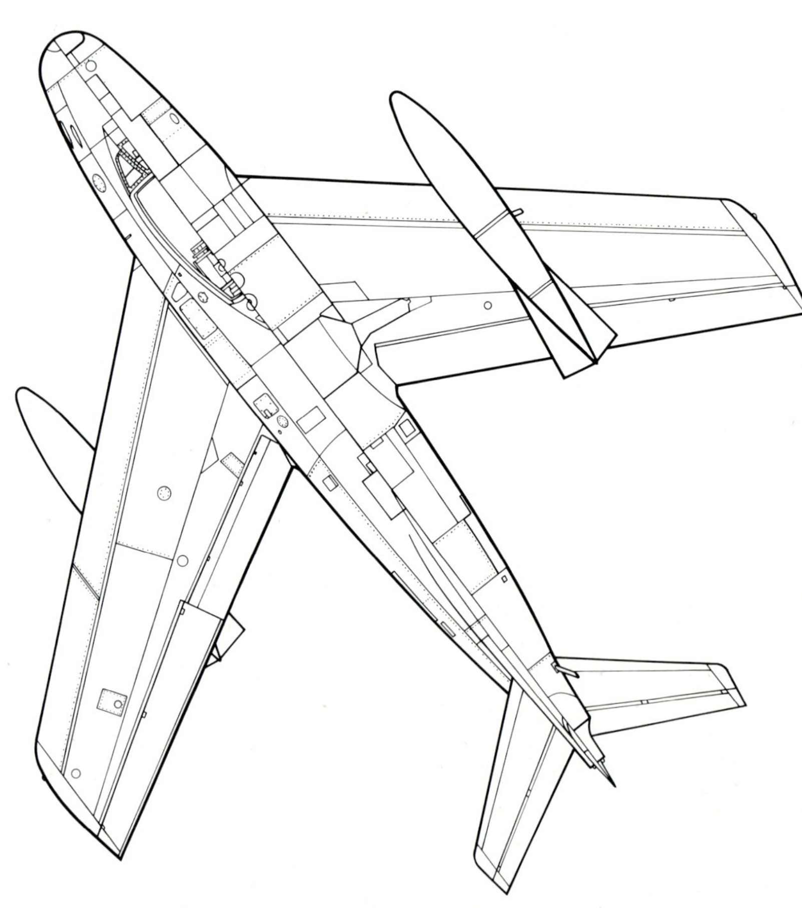

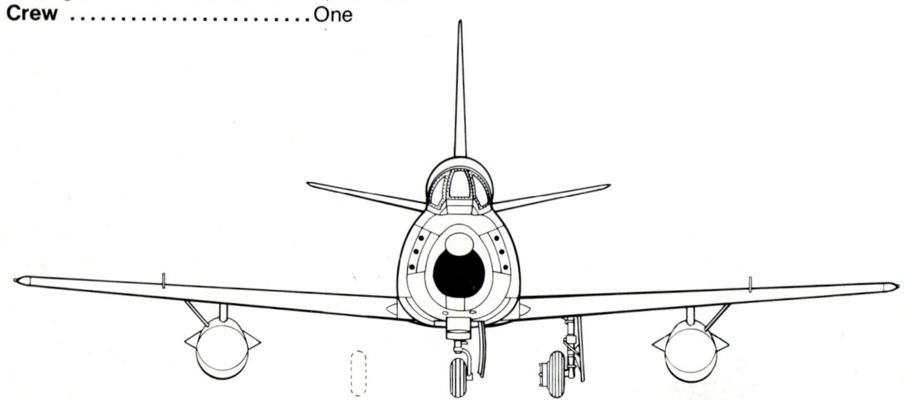

An F-86F of the 531st Fighter Bomber Squadron/21st Fighter Bomber Wing is rearmed with .50 caliber ammunition. The bomb in the foreground is a 500 pound GP bomb and will be loaded on the inboard wing pylon in place of the usual drop tank. (NAA)

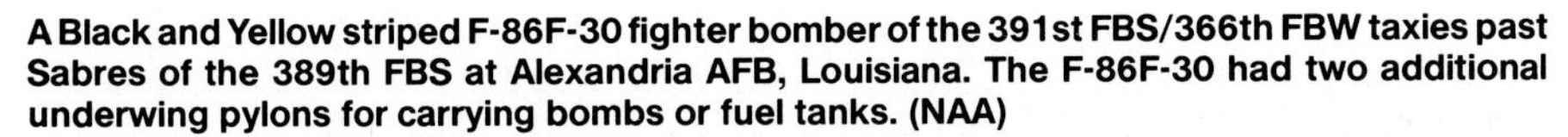

A Black and Yellow striped F-86F-30 fighter bomber of the 391st FBS/366th FBW taxies past Sabres of the 389th FBS at Alexandria AFB, Louisiana. The F-86F-30 had two additional underwing pylons for carrying bombs or fuel tanks. (NAA)

CAPT Joseph McConnell's famous F-86F-1 was named ***BEAUTIOUS BUTCH II***. McConnell was assigned to the 39th FIS and became the top scoring USAF ace in Korea with sixteen confirmed kills. For this later publicity photo, FEAF public affairs officials repainted the aircraft, removing the MiG silouette kill symbols, and misspelling the name (***BEAUTEOUS***). (NAA)

This Sabre Mk 5 was one of several Sabres obtained by the Honduran Air Force during the early 1970s. The aircraft was camouflaged in Dark Green/Medium Green/Tan uppersurfaces over Light Gray undersurfaces. The only markings was the Blue/White/Blue fin flash and the lettering FAH 3002 in Black on the fin. (COL Jim Bassett)

The F-86F-30 Sabre in the foreground was flown by MAJ Don Robinson when he commanded the 44th FBS. The aircraft were deployed to Thailand as part of Operation GOODWILL during June of 1954. The deployment was part of a show of force to the Viet Minh following the fall of French Indochina. (USAF)

F-86Fs flew alongside F-86Ds in the Air Defense Command during the Mid-1950s. This F-86F-25 was assigned to the 539th FIS at Stewart AFB, New York during 1954. The nose was Red with a White outline and the tail was Red with White stars. (COL J.P. Conti)

Silent George was an F-86F-10 of the 336th FIS, flown by CAPT George Love. The aircraft was about to depart Taegu for Tsuiki, Japan for retrofit with the hard edge wing. NAA sent some fifty wing kits to Japan for installation on F-86Fs flown by aces. (Alan Fine)

This F-86F was flown by the Washington D.C. Air National Guard aerobatic team, the ***MINUTEMAN***. Very few F-86Fs were transferred to the Air National Guard: most were sold under the Military Assistance Program to foreign air forces. (Dave McLaren)

Armorers prepare to load a M117 1,000 pound GP bomb on the inboard pylon of *The Georgia Peach,* an F-86F-30 of the 36th Fighter Bomber Squadron/8th Fighter Bomber Wing at K-13, Korea during June of 1953. (USAF)

An F-86F-30 flown by the commander of the 8th FBW, COL Walter Benz is displayed next to an F-86F-10 flown by COL George Ruddell. The F-30 is carrying "Misawa" drop tanks (made in Japan) while the F-10 is carrying the original NAA combat drop tanks. (NAA)

The 18th Fighter Bomber Wing deployed to Korea with F-86F-30s during early 1953. The inboard pylon was stressed for loads up to 1,000 pounds and was plumbed for carrying an additional fuel tank. This is an early F-30 with the original short chord wing with LES. (NAA)

MY GEORGIA **an F-86F of the 16th FIS stands strip alert at Suwon during the Winter of 1953. Although the Korean War was over, there was still contact and combat between USAF Sabres and North Korean MiG-15s and MiG-17s through 1955. (Dick Geiger)**

An RCAF Sabre Mk 5 of No 414 Squadron shares the ramp at Dharan, Saudi Arabia with an F-86F of the 48th FBW during Operation MORNING STAR held in January of 1955. Thirty-six years later, USAF and Canadian Forces fighters would once again share the ramp at Dharan during Operation DESERT STORM. (USAF)

This F-86F was flown by COL Spicer, commander of Air Training Command. Each stripe on the fuselage and tail was a different color with the name of an Air Training Command base being carried on the stripe against a White background. (Enger via Marty Isham)

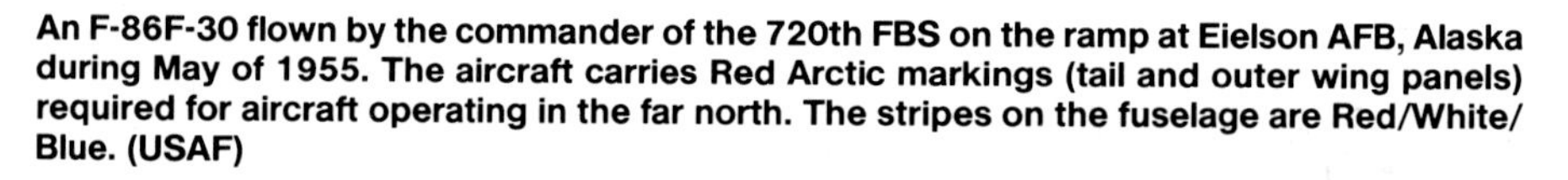

An F-86F-30 flown by the commander of the 720th FBS on the ramp at Eielson AFB, Alaska during May of 1955. The aircraft carries Red Arctic markings (tail and outer wing panels) required for aircraft operating in the far north. The stripes on the fuselage are Red/White/ Blue. (USAF)

F-86F Sabre fighter-bombers of No 2 Squadron, South African Air Force taxi to the active runway at Osan AB (K-55) during the Spring of 1953. No 2 Squadron returned their Sabres after Korea and purchased Canadair variants to supplement the Sabre force in South Africa. (NAA)

During the mid-1950s, several USAF units experimented with NATO style camouflage on both F-86s and F-84s. This F-86F of the 461st FDS is camouflaged with Medium Gray and Dark green uppersurfaces with British PRU Blue undersurfaces. The project was later cancelled and the paint stripped off. (Rob Satterfield)

An RCAF Sabre Mk 6 of No 422 Squadron at Baden-Solligen during 1962. The Mk 6 differed from earlier Sabre variants in having a 7,275 lbst Orenda 14 engine. The Mk 6 was the last Canadair Sabre produced and was widely exported. (Brian Baker)

This Canadair Sabre Mk 6 was flown by the Columbian Air Force. The aircraft carried standard NATO Dark Green/Dark Gray camouflage. The rudder stripes were (top-bottom) Yellow/Blue/Red, all lettering was in Black. (Robb Satterfield)

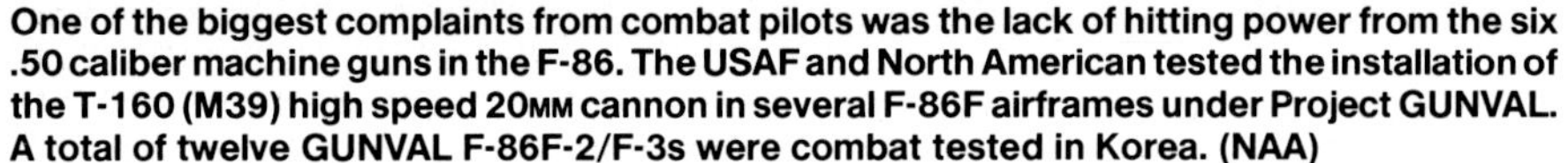
One of the biggest complaints from combat pilots was the lack of hitting power from the six .50 caliber machine guns in the F-86. The USAF and North American tested the installation of the T-160 (M39) high speed 20MM cannon in several F-86F airframes under Project GUNVAL. A total of twelve GUNVAL F-86F-2/F-3s were combat tested in Korea. (NAA)

This F-86F was "bailed" (loaned) to the NASA facility at Ames Laboratory where it was used as both a chase aircraft and for high speed flight tests. The markings were Dark Blue with Yellow trim. (Mike O'Conner)

Project GUNVAL

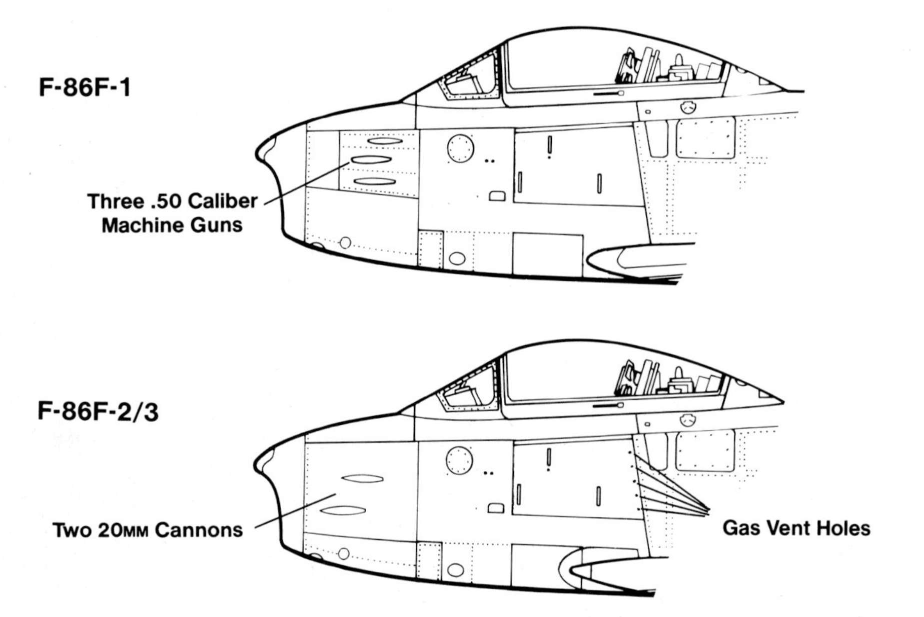

Japan was one of the largest users of the Sabre, flying them well into the 1980s. Mitsubishi Heavy Industries built 300 Sabres under the designation F-86F-40s. The F-40 wing was extended twelve inches at each wingtip and had the extended chord wing leading edge fitted with leading edge slats. (Jim Sullivan)

The Royal Saudi Air Force was equipped with F-86F fighter bombers taken from USAFE inventories and fitted with the F-40 wing. U.S. maintenance instructors show a Saudi pilot and crew chief the workings of the Sabre cockpit at Riyadh in November of 1966. The aircraft was overall Light Gray with Green lettering. (Frank McDonald)

NATO air forces were very satisfied with the Sabre, regardless if it was North American or Canadair built and both were operated within NATO during the 1950s. This NAA-built F-86F-35 is assigned to No 331 Fighter Squadron, Norwegian Air Force during February of 1961. (Merle Olmsted)

The Republic of Korea Air Force (ROKAF) was given a number of Sabres following the Korean War, flying them into the 1980s. This ex-USAF F-86F-25 was retrofitted with the 'F-40' extended wing and modified to carry the AIM-9 Sidewinder air-to-air missile. (Robert C. Mikesh)

Late in their careers, Republic of Korea Air Force (ROKAF) F-86F-25 Sabres were camouflaged with Dark Green and Tan uppersurfaces over Light Gray undersurfaces. All lettering was in White. (Centurion)

Although pioneered for use by USAF units in Korea, the primary operators of the RF-86F were the Japanese and South Korean air forces. This ROKAF RF-86F is camouflaged in very faded Dark Green and Tan uppersurfaces with Light Gray undersurfaces. This RF-86F carried three cameras in the pack and no guns. (Jim Sullivan)

RF-86F

F-86F-40

Six .50 Caliber Machine Guns

RF-86F-40

Gun Ports Faired Over (Some Aircraft)

Bulged Equipment Bay Door

Camera Fairings

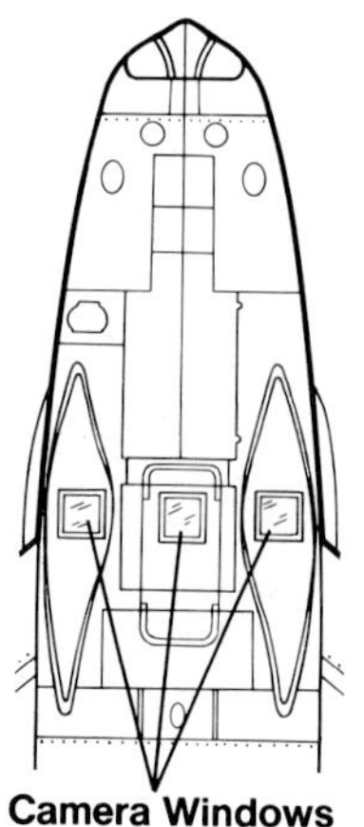

A Japanese Air Self Defense Force (JASDF) RF-86F takes off for a training mission in Japan. This RF-86 retains the gun ports for the six .50 caliber machine guns. The aircraft was built in Japan under license by Mitsubshi. (N.J. Waters III)

In 1953 the USAF expressed an interest in a trainer version of the Sabre. North American engineers pulled an F-86F-30 from the production line and stretched the fuselage sixty-three inches to make room for the second cockpit and moved the wing forward eight inches to restore the center of gravity. (NAA)

Fuselage Development

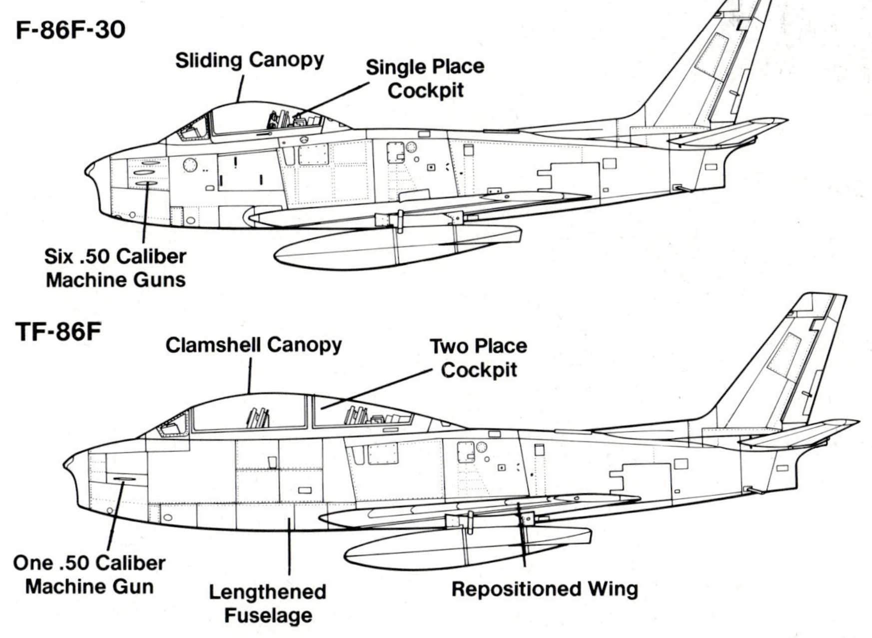

The TF-86F did not see much service in the mission for which it had been designed. The second TF-86F prototype, however, did see extensive duty at Edwards AFB as a chase aircraft where it served well into the 1960s. (Brian Baker)

Commonwealth CA-26/27 Sabre

The most radical license-built variant of the Sabre was the Australian Commonwealth Aircraft Corporation CA-26. The CA-26 mated an F-86F airframe with a 7,500 lbst Rolls Royce RA 7 Avon engine. Sounds simple enough but the installation required a number of airframe changes. Since the Avon engine consumed some twenty-five percent more air than the J47 engine, Commonwealth engineers had to split the forward fuselage and splice in an additional three inches to the air intake area. Although of a similar size dimensionally as the J47, the Avon was some 400 pounds lighter. This meant that the aircraft's center of gravity would be thrown off. To preserve the standard Sabre c/g, the Avon engine was positioned further back in the fuselage. Moving the engine also meant that the fuselage "break point" (where the forward and rear fuselage sections split for engine servicing) also had to be moved rearward.

The armament was also changed with the six .50 caliber machine guns being replaced by a pair of Aden 30MM cannons. The fuselage fuel cells and accessories were also relocated. By the time Commonwealth engineers had finished with the engine installation, only twenty-five percent of the original North American fuselage remained. The wing and tail section were retained from the F-86F, including the "all-flying tail," and the narrow chord wing with LES. The CA-26 prototype (A94-101) was completed in early July of 1953 and RAAF FLT LT W. Scott took the Sabre Mk 30 into the air for the first time on 3 August 1953.

Production aircraft were designated the CA-27 and received serials beginning with A94-901. The first CA-27/Mk 30 production aircraft was delivered on 19 August 1954 and following production of the initial batch, North American began shipping wings to CAC equipped with the '6-3' leading edge extension and wing fences. CA-27 Sabres with the '6-3' wing were designated as Sabre Mk 31s. All previous Mk 30 aircraft were later retrofitted with the '6-3' wing as they went through IRAN overhauls. The Mk 32 used the new CAC-built Mk 26 Avon engine and had provisions for underwing ordnance in addition to a pair of drop tanks (similar to the USAF F-86F-25/30). In 1959, the RAAF began fitting Sidewinder missile launch rails on these additional hard points. CAC built a total of 112 Avon Sabres between August 1954 and December 1961.

The first operational Avon Sabre unit was No 75 Squadron commissioned in April of 1955. The RAAF had a total of six Sabre units, five fighter squadrons and an operational conversion unit (OCU). RAAF Sabre units saw combat in the attempted communist takeover in Malaya in the late 1950s. Nos 3 and 77 Squadrons flew ground attack and counter-insurgency missions beginning in February of 1959, continuing until July of 1960. RAAF Avon Sabres were on a combat footing again beginning in 1962 when No 79 Squadron began an extended stay at Ubon RTAFB in Thailand. They would provide combat air patrols over Thailand during the early stages of the Vietnam War. As RAAF units converted to higher performance aircraft such as the Dassault Mirage III, the Avon Sabres were made available for sale. Malaysia received eighteen Avon Sabres during 1969 and 1971. Indonesia purchased eighteen Avon Sabres from the RAAF in February of 1973 and another five from Malaysia during July 1976.

The Australian Commonwealth Aircraft Corporation combined the F-86F airframe with the 7,500 lbst Rolls Royce Avon 26 engine to produce the CA-26/27 Sabre series. The CA-26 prototype made its first flight on 3 August 1953. (Jeffery Ethell)

A Commonwealth CA-27 Sabre Mk 32 of 76 Black Panther Squadron, RAAF. The unit had formed an aerobatic team which flew the Sabre during the 1960s. The Mk 32 had both the '6-3' wing with fences and Sidewinder missile capability. RAAF Sabres flew combat air patrols over Thailand during the very early stages of the Vietnam War. (Brian Baker)

Fuselage Development

F-86F-30

Fuselage Break

Three .50 Caliber Machine Guns

CA-27

Additional Air Intakes

Repositioned Fuselage Break

Deeper Intake

30MM ADEN Cannon

Following their retirement from service with the RAAF, the Malaysian and Indonesian governments each purchased CA-27 Sabre Mk 32s from RAAF inventories. This Malaysian Air Force Sabre Mk 32 was camouflaged in Dark Green over Light Gray.

F-86H

The last Sabre variant built by North American was the F-86H and it was intended to be the best of the Sabre line. All the previous problems and "needs" were going to be answered with the F-86H. The additional power that Sabre pilots craved was answered by installing a 8,900 lbst General Electric J73 engine, which required a six inch increase in fuselage depth. The lack of fire power was greatly improved by replacing the six .50 caliber machine guns with four of the newly developed M39 20MM high speed cannons (although early production F-86H-1s retained the six .50 caliber gun armament). The F-86H initially had the '6-3' wing with fences, but was later fitted with the F-40 extended wing and eventually all production F-86Hs were retrofitted with the F-40 wing. The F-86H was equipped with the LABS system and could carry a 1,200 pound "special store" (atomic weapon) on the inboard underwing hard points.

Unfortunately, the airframe itself had reached its limit and no amount of additional power could bring the F-86H into the level supersonic speed range. That fact, plus development of the "Sabre 45" (F-100 Super Sabre), led to a short production run of only 473 aircraft (at a flyaway cost of $582,839.00 each). Although both of the prototypes (52-1975 and 52-1976) were built at the Inglewood North American plant, all production aircraft were built at the North American plant in Columbus, Ohio. The first production aircraft were delivered to the 312th FBW at Clovis AFB, NM in the Fall of 1954 while the last F-86H was delivered to the Air Force in October of 1955. F-86Hs equipped five USAF fighter-bomber wings in the mid-1950s.

The first F-86H aircraft were already being phased into Air National Guard service by 1957 and by June of 1958, all F-86Hs in active Air Force use had been passed to the ANG, although there was a brief time when a number of the F-86Hs were returned to active duty. During the 1961 Berlin Crisis, the Air Force activated the 101st and 131st Tactical Fighter Squadrons, Massachusetts Air National Guard, deploying them to Phalsbourg Air Base, France where they remained until August of 1962.

It was an F-86H that flew the final USAF/ANG Sabre sortie, when the 138th TFS, New York ANG phased the Sabre out of service on 30 September 1970. Following phaseout with the Air National Guard, the F-86Hs with the lowest flight hours were made available to the U.S. Navy. The Navy had a dual use for the Sabres — as a target drone and as a MiG simulator in air combat training. F-86Hs were used in the Navy TOP GUN program as an aggressor aircraft since they were of a similar size, shape and performance as the MiG-17s being encountered by Navy pilots over North Vietnam. More than one Navy TOP GUN F-4 pilot was "killed" by a Sabre. Finally, the F-86Hs were used as target drones for testing advanced air-to-air and surface-to-air systems such as the Phoenix, AMRAAM, and Standard missiles.

North American Chief Test Pilot "Wheaties" Welch stands by the nose of the F-86H prototype (52-1975). The F-86H was to have solved all the complaints about the Sabre program, since it had much more power and a cannon armament. (NAA)

The YF-86H-1 service test aircraft at Edwards AFB during 1954. The early production F-86H-1s retained the six .50 caliber machine gun armament, which was replaced by M39 cannons beginning with the 117th aircraft off the production line. (USAF)

This F-86H-1 was bailed to Lockheed for use as a chase aircraft for the F-104S and A-12 programs during the late 1950s and early 1960s. The nose, tail and drop tank colors are Red with the Yellow Lockheed "star" marking. (JEM Slides)

An F-86H-1 of the 312th FBW based at Clovis AFB, New Mexico during 1954. Initial production aircraft had the '6-3' wing with fences, which was later replaced by the F-40 wing with slats. The F-86H was atomic capable and could toss bomb the 1,200 pound "special store" using the LABS system. (COL E.M. Hanley)

Following a tour in Korea in F-84 Thunderjets, the 474th FBW rotated home to Clovis AFB and transitioned to the F-86H. This 428th FBS F-86H-5 has Black and Yellow bands on the nose, tail, fuselage, wingtips and drop tanks along with a Day-glo Red 5 indicating it was assigned to a member of the 474th FBW gunnery team. (Robb Satterfield)

Fuselage Development

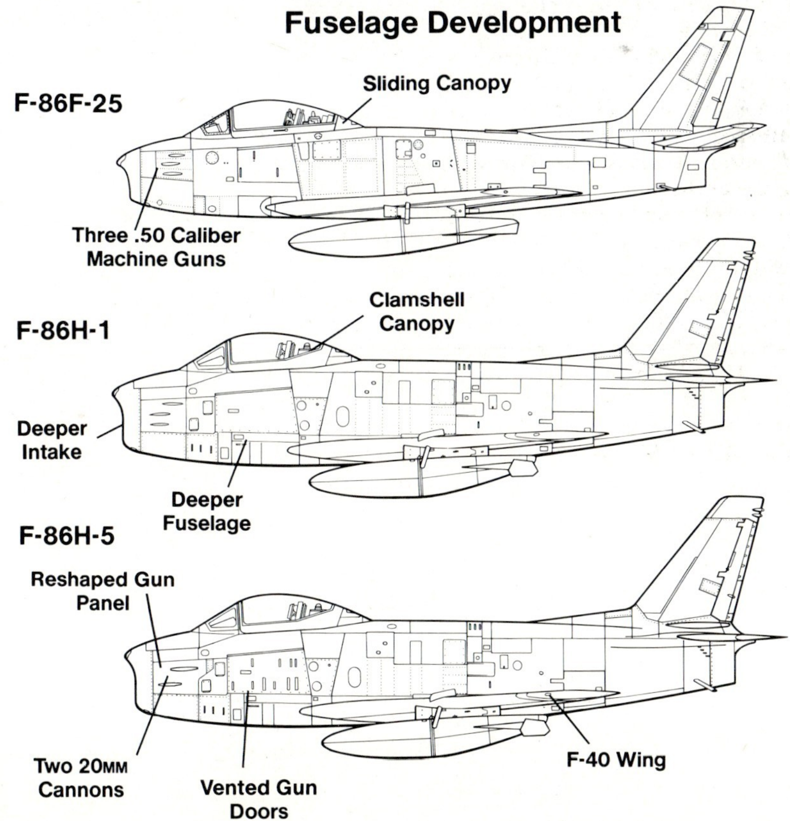

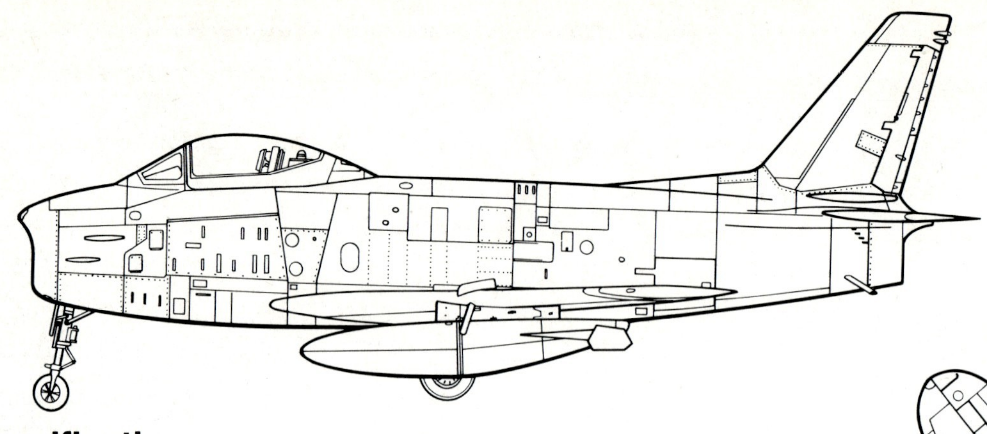

Specifications

North American Aviation F-86H-5 Sabre

Wingspan .... 38.84 feet
Length .... 39.12 feet
Height .... 14.99 feet
Empty Weight .... 13,836 pounds
Maximum Weight .... 21,800 pounds
Powerplants .... One 8,920 lbst General Electric J73-GE-3D turbojet engine

Armament .... Four M39 cannons and up to 2,400 pounds of underwing stores.

Performance
Maximum Speed .... 692 mph
Service ceiling .... 50,800 feet
Range .... 1,810 miles
Crew .... One

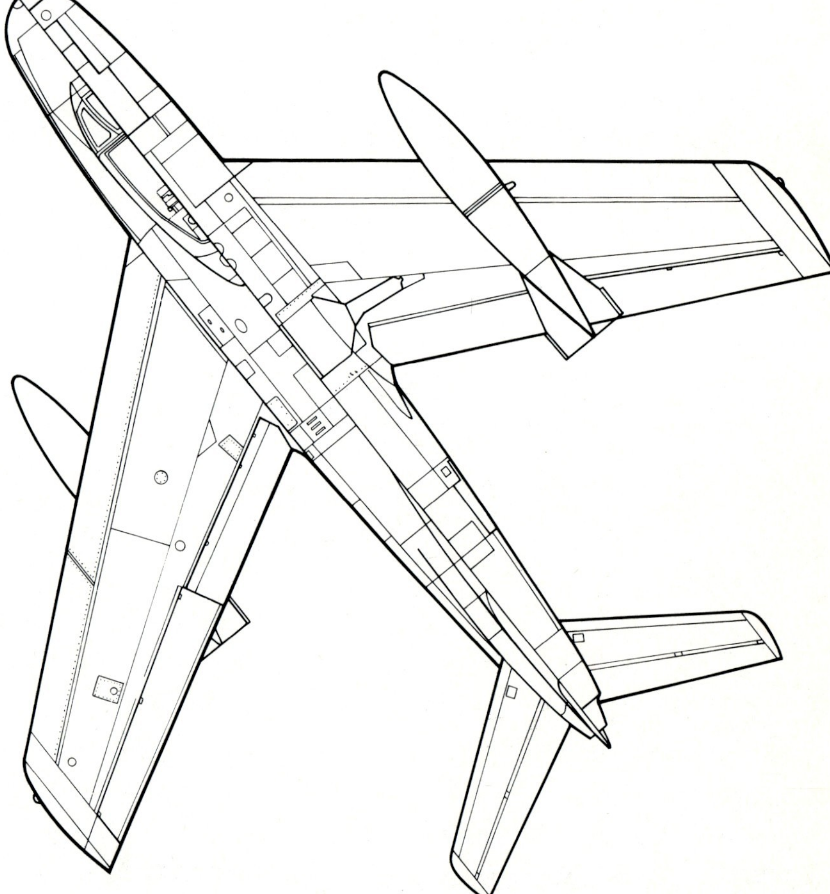

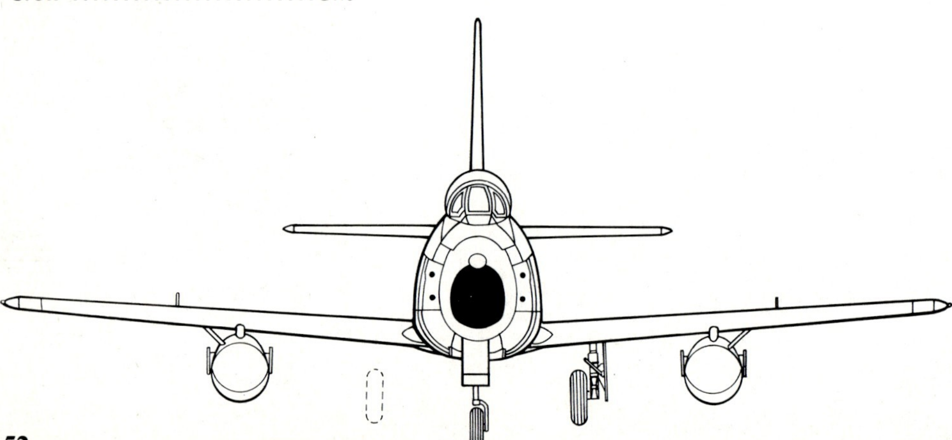

Ground crewmen work on a completely stripped F-86H of the 474th Fighter Bomber Wing on the open ramp at Clovis AFB. The J73-GE-3 engine was much larger than the J47 and was rated at 8,900 pounds dry thrust. Like the earlier F-86Ds, the F-86H used a clamshell type canopy. (NAA)

This highly polished F-86H-10 was flown by COL Fred Ascani when he commanded the 50th Fighter Bomber Wing at Hahn Air Base, West Germany during 1956. The F-86H-10 had both the M39 cannon armament and the F-40 extended wing. (MGEN Fred Ascani)

JP-4 fuel tankers refuel a trio of 428th FBS/474th FBW F-86H-5s at Clovis AFB during 1954. Repeated problems with the M39 cannon installation resulted in cracked gun bay walls and even fuselage cracks. This, plus developmental problems with the J73 engine, brought a quick end to the USAF service life of the F-86H. (NAA)

An F-86H-10 of the 10th FBS/50th FBW at Hahn AB, West Germany in the mid-1950s. The aircraft is configured for long range flight with four underwing drop tanks. Although certainly a good aircraft, the F-86H simply came too late and was quickly replaced by the North American supersonic F-100 Super Sabre. (via David Menard)

A flight of 121st Tactical Fighter Squadron F-86Hs based at Andrews AFB in flight near the nation's capitol. The 121st TFS exchanged their F-86Hs for F-100Cs during 1960. The F-86H had vents in the gun bay doors to relieve gun gas buildup. (David Menard)

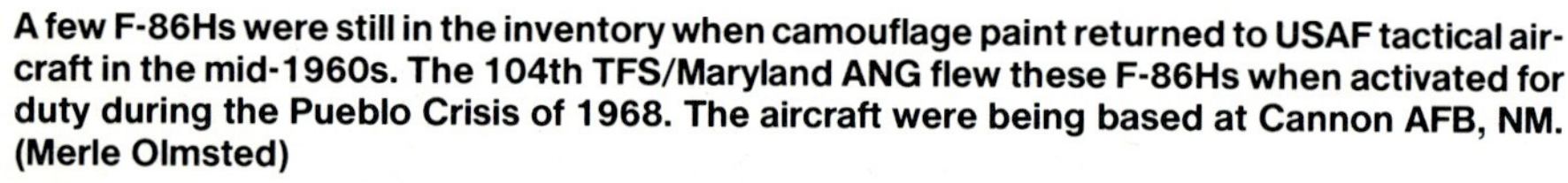

A few F-86Hs were still in the inventory when camouflage paint returned to USAF tactical aircraft in the mid-1960s. The 104th TFS/Maryland ANG flew these F-86Hs when activated for duty during the Pueblo Crisis of 1968. The aircraft were being based at Cannon AFB, NM. (Merle Olmsted)

F-86Hs were rapidly phased into service with Air Guard units beginning in 1956 as F-100s became available for active USAF units. These F-86H-10s are assigned to the 195th FIS/California ANG. These aircraft have been painted overall Silver lacquer for corrosion control. (NAA)

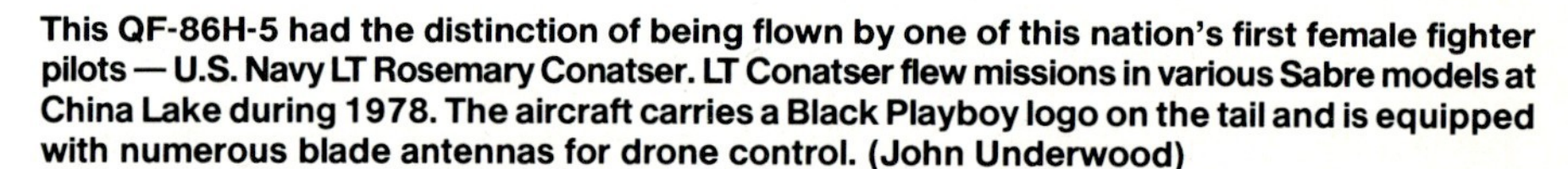

This QF-86H-5 had the distinction of being flown by one of this nation's first female fighter pilots — U.S. Navy LT Rosemary Conatser. LT Conatser flew missions in various Sabre models at China Lake during 1978. The aircraft carries a Black Playboy logo on the tail and is equipped with numerous blade antennas for drone control. (John Underwood)

F-86K

Another major development in the Sabre line was the Fiat of Italy license-built F-86K. The F-86K model was the answer to a NATO requirement for an all-weather jet interceptor to counter the Soviet nuclear bomber threat. Basically the F-86K was an F-86D simplified for use by NATO and other friendly foreign air forces. The F-86K was armed with four M24A 20MM cannons in place of the all-rocket armament of the F-86D. The F-86K also used an MG-4 Fire Control System with an optical gunsight in place of the much more sophisticated Hughes E-4 system found on USAF aircraft and the F-86K retained the AN/APG-36 search radar used with the F-86D.

Initially, the F-86K was to have followed all the engine changes found in the F-86D program. The Air Force, however, wisely decided to use a single power plant, the J47-GE-33 for all the F-86Ks, simplifying production. As with the F-86D and F-86L, the F-86K had the early narrow chord wing with leading edge slats, but was later retrofitted with the F-40's extended wing. Finally, many F-86Ks had their armament upgraded to include a pair of Sidewinder air-to-air missile in addition to their four cannons.

The first flight of the YF-86K prototype took place on 15 July 1954. Both YF-86Ks were converted F-86D-40 airframes (52-3630 and 52-3804). Following service tests at Edwards AFB, both YF-86Ks were delivered to Fiat in Italy. North American built 120 F-86Ks at the Inglewood plant, which were then disassembled and shipped to Fiat for assembly, while Fiat built another 221 F-86Ks.

The first F-86Ks went into service with the 1st *Aerobrigata*, Italian Air Force in the late Summer of 1955. NATO air forces that operated the F-86K included the Italian, French, Federal German, Dutch and Turkish air forces. Phaseout of the F-86K began during 1964 when Italian Air Force units began to transition to the Lockheed F-104 Starfighter.

This Italian Air Force F-86K was part of the initial production batch of Fiat-built aircraft. Fiat built a total of 221 F-86Ks while North American built an additional 122. This aircraft has both the F-40 wing and Sidewinder missile rails. (IAF)

The F-86K was the North American Aviation answer to a NATO requirement for a gun armed all-weather interceptor. Basically the F-86K was an F-86D-45, with a less sophisticated MG4 Fire Control System and four cannon armament. (USAF)

Armament Development

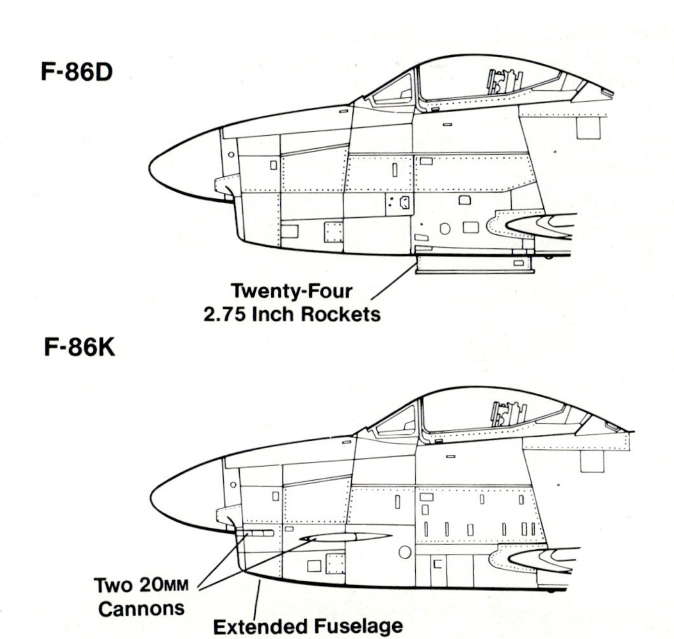

An F-86K-NF of *13 Esquadre*, French Air Force during 1958. Initially F-86Ks were delivered with the standard, short-chord, short-span F-86D wing and later retrofitted with the F-40 wing. France purchased a total of sixty Fiat-built F-86Ks. (French Air Force)

A West German *Bundesluftwaffe* Fiat-built F-86K of JG-74 during 1967. West German F-86Ks were painted the typical NATO tactical camouflage of Dark Green and Dark Gray uppersurfaces over Light Gray undersurfaces. (Manfred Franzke)

The Honduran Air Force acquired several F-86Ks during the 1960s. Sources have disagreed as to the source of these aircraft, some state that they were ex-West German via Venezuela while other say the aircraft were ex-Yugoslav and still others say the aircraft were ex-Danish. It is now known that they were ex-German. (COL Jim Bassett)

This overall Light Gray Venezuelan Air Force F-86K was one of a number of ex-German F-86Ks purchased by Venezuela during the late 1960s. Only half of the aircraft purchased were put into service, the others being used as a source of spare parts. Later several of these crated aircraft were passed to Honduras.

Korea

Any conversation regarding the Sabre automatically leads to the subject of Korea, for it was over Korea that the Sabre proved its mettle. What makes the topic even more interesting is that in almost every aspect of air combat with the MiG-15, the Sabre was never its equal — yet the results of that Police Action revealed at least a 10-1 kill ratio in favor of the Sabre.

A comparison of the performance characteristics of the MiG-15 and the Sabre are extraordinary when viewed in the results of the war. The MiG was much smaller and lighter, weighing slightly over 11,000 pounds, compared with the F-86E at 19,000 pounds. The MiG was powered by a Soviet copy of the Rolls Royce Nene engine, the 5,000 lbst RD-45, while the F-86F had the most powerful Sabre engine with 5,910 lbst. The MiG and Sabre were roughly comparable in top speed at both sea level and at combat altitudes of 40,000 feet or above. But the lighter weight of the MiG gave it a phenomenal rate of climb of over 10,000 ft/min. The F-86F with the -27 engine could muster 9,300 ft/min; however, the Sabres that met and defeated the MiGs in 1950-52 were mostly F-86As and F-86Es, both of which had a rate of climb of only 7,400 ft/min.

The lightweight MiG also had about a 7,000 foot advantage in service ceiling. Sabres in Korea were entering MiG Alley at 45,000 feet only to find the MiGs were about 10,000 feet above them — and there was nothing the Sabre pilots could do about it unless the MiGs came down. If they did come down, their advantages started fading rapidly. The Sabres being much heavier than the MiGs, had a much higher diving speed. Both the Sabre and MiG could go supersonic in a dive, but the Sabre was a much more stable aircraft in the transonic speed range. The MiG-15 became quite unstable when entering the transonic speeds near Mach one. It was not uncommon for a Sabre pilot to push a MiG pilot into the transonic speed area only to see the MiG lose the entire vertical tail assembly during violent combat maneuvering. The MiG was a much more maneuverable aircraft than the F-86A, F-86E and early F-86F. But the later F-86F with the hard wing leading edge and wing fences erased that advantage.

Armament was a tie with neither having a clear advantage. The MiG had much greater hitting power with its twin 23MM cannons and single 37MM cannon. But the cyclic rate of these guns was very slow and their ammunition supply was low: 160 rounds per gun for the 23MM guns and forty rounds for the 37MM gun. The F-86s were armed with six .50 caliber machine guns. The machine guns had a much higher cyclic rate, could carry much more ammunition, but did not have the stopping power of the cannon. It was not unusual for a Sabre pilot to empty all 1,600 rounds into a MiG, only to have the MiG escape back over the Yalu. The Sabre did have a radar-ranging gunsight, which was far more accurate than the older gyroscopic sight in the MiG. This accuracy, however, was wasted due to the low weight of fire from the six .50s. Add to all this the fact that the Sabre flights were always outnumbered and were flying at the extreme end of their range. Communist MiG strength ran between 400 and 500 MIGs in early 1951, to over 1,000 MiGs by the end of the war. This was opposed by a single squadron of F-86As in 1951, to six squadrons in early 1953 and finally eleven squadrons at the end of the war.

So how did we end up with that extraordinary kill ratio? The answer is simple — the pilots! The pilots of the 5th Air Force were much better trained than the best the Communists sent to Korea. Most of the Sabre pilots, especially the squadron and group commanders, had extensive combat experience in the Second World War. Although the Russian pilots that flew in Korea had some combat experience, most of the North Korean and Chinese pilots had none. All this Second World War experience meant that the Sabre pilots had already developed and proven the combat tactics they would be using in Korea. Tactics that had worked against a much better foe six years earlier.

The most important ingredient that brought success to the Sabre pilots was their aggressiveness in combat. Where a MiG pilot might be wary of combat, the Sabre pilots were looking for it. As the Navy ads say, "It's not just a job." Always outnumbered in MiG Alley, Sabre pilots never hesitated to join combat with the MiGs — no matter what the odds were. One Sabre pilot once exclaimed over the radio that, "I have 24 MiGs cornered near Sinanju in case anyone isn't busy!" It was that type of aggressive attitude that gave the combat edge to the Sabre pilots. Add to all this the technological advances made to the Sabre airframe and systems, advances that the Sabre pilots always took full advantage of, and one can now begin to see how the 10-1 victory margin came about.

The seven F-86As that LCOL Bruce Hinton took to K-14 on 13 December 1950 all had the short-chord wing with LES, a gyroscopic gunsight, standard tail and the 5,200 lbst -13 engine. Yet they wrestled control of the skies over North Korea from the MiGs in one afternoon. By the time the F-86F-25 arrived with its improved wing, radar-ranging gunsight and more powerful engine, Sabre pilots had total air supremacy all the way to the Yalu River and beyond! At the end of the war, the final score was amazing. The final score was 792 MiGs shot down by F-86s, for a loss of 78 Sabres to the MiGs — a 10-1 victory margin. It was the greatest victory margin of any U.S.-involved conflict, excluding the latest confrontation in the Gulf. The phenomenal victory margin was the result of putting a superbly trained, aggressive pilot into a superbly designed and developed fighter aircraft, using combat tactics developed by proven combat leaders. Although the war in Korea officially ended as a tie, the skies belonged to the Sabre Pilots!

CAPT Jabara's F-86 on the ramp at Suwon the day after his twin MiG victories. The aircraft carried no personal markings or victory markings. Jabara scored the twin kill on 20 May despite having a hung drop tank. (Leo Fournier)

During the Summer of 1951 several F-86As, RF-80s and AT-6s had their upper surfaces painted Olive Drab in an attempt to camouflage them from the higher flying MiGs. The experiment failed miserably and the painted Sabres actually stood out more than a natural metal aircraft. The paint cost some 20 mph in top speed. (Elflein)

LITTLE BUTCH, an F-86A of the 335th FIS, carries the markings adopted by the 4th FIW during 1952. The Black and Yellow FEAF ID bands were carried on the wings and fuselage, with another Black/Yellow band on the tail to denote an aircraft assigned to the 4th. (Karl Dittmers)

LT Cecil Foster flew this F-86E while assigned to the 25th Fighter Interceptor Squadron at K-13 airfield, Korea during 1952. LT Foster scored his fifth MiG kill on 22 November to become the twenty-third ace of the Korean War. (NAA)

Bonnie and Cuddles, an F-86E of the 334th FIS, 4th FIG taxies out to the active runway at Kimpo for another mission. The 4th FIW accounted for some 502 MiGs during the two and one half years they were in combat in Korea. When the war ended, the skies over Korea belonged to the Sabres.(NAA)